PREFACE

This report provides, for each of the 54 states and territories, the identity and responsibilities of the directors of the following offices: state emergency management agency; state homeland security agency; and state adjutant general and/or military department. In states where the functions of homeland security and emergency management are combined, this is so indicated. The report also enumerates the qualifications necessary for the director position and indicates the statutory or regulatory authority under which the position exists. For each state, the researchers have retained the language used in the respective state codes and/or statutes.

The authors of this report researched three primary online sources: state codes and statutes; executive orders; and Web sites of the respective agencies. In instances where laws could not be identified online, the resources of the Law Library of Congress were used. Finally, for those states/territories that did not provide the requisite information online, the researchers telephoned the relevant homeland security or emergency management office.

I0393848

ii

TABLE OF CONTENTS

OVERVIEW

In general, the states of the United States and its territories share much in common regarding the functions and responsibilities of the offices of emergency management, homeland security, and the state military. As table 1 indicates, more than half of the states (27) and territories (3) have some degree of jurisdictional overlap among emergency management, homeland security, and adjutant general departments. Of the 55 states and territories, 22 have merged emergency management and homeland security functions into one department. In four states that have separate emergency management and homeland security departments, there is a common director. In some states, the adjutant general's functions incorporate those that are generally assigned to homeland security and/or emergency management directors. In eight of the 55 states and territories, the adjutant general heads the emergency management or homeland security department.

Common Functions/Responsibilities of Homeland Security and Emergency Management Departments

Directors of emergency management and homeland security are generally tasked to prepare and maintain a comprehensive plan and program for emergency management or homeland security. This usually entails the implementation and administration of a statewide strategy for emergency management and homeland security. State directors also coordinate the emergency and homeland security activities of all state agencies, and coordinate with the emergency management/homeland security plans of the federal government and other states. Some states, like Florida, extend the coordination function to county and municipal governments, school boards, and private agencies that have an emergency management/homeland security role. Directors also coordinate the distribution of information and security warnings to state and local government and the public.

With regard to disaster and terrorist incident response, it is also the responsibility of directors to develop policies to train local, regional, and state officials in proper procedure. The terminology used to define the directors' role in preventing and responding to both natural disasters and attack threats to persons and critical infrastructure in the various states and territories has a common thread. These entities share a common need to prevent and respond to acts of terrorism and other critical hazards. The key words used in state and territorial statutes and agency mission statements are: preparedness and training; prevention; prompt, effective emergency response and recovery; minimization of injury; and identification of areas vulnerable to disaster and emergency. A general directive for state directors could be summed up by the words *detect*, *deter*, *mitigate*, and *respond*.

Overlapping Functions

In some states, the overlap between the emergency management and homeland security functions is evident in the allocation of responsibilities to the directors, or the chain of command, rather than in the bureaucratic structure of the departments. States that fit this category are:

Alabama – Emergency management and homeland security are separate departments, but the director of emergency management is also the Assistant Director of Homeland Security for Emergency Preparedness and Response. The director is required by statute to "maintain liaison with and cooperate with major commanders of the armed forces within the state, and the State Military Department."

Arizona – The director of emergency management is appointed by the adjutant general, and his or her responsibilities, as defined by statute, are subject to the approval of the adjutant general.

Idaho – The Bureau of Homeland Security and Disaster Emergencies is headed by a chief appointed by the adjutant general with the concurrence of the governor; the governor can also opt to appoint the adjutant general as chief of the bureau. The adjutant general serves as the governor's authorized representative for emergency planning, preparedness, response, and recovery from all hazards.

Kentucky and Maryland – The positions of adjutant general and director of emergency management are held by two different persons, and their respective departments are separate; however, the emergency management director is accountable to the adjutant general.

Maine – The adjutant general is the commissioner of the Defense, Veterans and Emergency Management Department. He also serves as the governor's official homeland security adviser.

Minnesota – The director of the Division of Homeland Security and Emergency Management is within the state's Department of Public Safety. Under statute, only the functions and responsibilities of the State's Division of Emergency Management are enumerated.

Missouri – The State Emergency Management Agency (SEMA) was created under statute "within the military division of the executive department, office of the adjutant general;" the adjutant general is the executive head of SEMA. Currently, a civilian is director of the agency.

Montana – The Division of Emergency Services (DES) is within the state's Department of Military Affairs. DES serves as the lead agency for the state's Homeland Security Task Force.

State Police Functions

Some states have opted to assign emergency management and/or homeland security functions to their law enforcement/state police departments. In Florida, the director of the state's Department of Law Enforcement, working closely with the Division of Emergency Management, is tasked to coordinate the state's detection, prevention, preparation for, response to, and recovery from, acts of terrorism. In Michigan, the Director of State Police also serves as State Director of Emergency Management and as the state's Homeland Security Director. In New Jersey, the State Police directs emergency management, and its Homeland Security Branch is tasked to "provide a continuing preventive level of homeland security and public safety through the coordination of statewide resources." In 2006 the governor created an Office of Homeland Security and Preparedness to administer and supervise the state's counterterrorism and preparedness efforts. Recognizing the potential overlap in functions between the state police and the newly created homeland security office, the governor's executive order stipulates that although the state police will continue to operate the Office of Emergency Management, the police superintendent is to provide dual reporting to the attorney general and the director of the homeland security office on matters related to homeland security, preparedness, and the emergency management office.

Table 1: Functions of State Directors

State/Territory	Combined EM/HS*	Separate EM & HS but Same Director	Adjutant General Directs EM and/or HS
Alabama			
Alaska	X		
Arizona			
Arkansas	X		
California			
Colorado		X	
Connecticut	X		
Delaware			
Florida			
Georgia		X	
Hawaii	X		X
Idaho	X		
Illinois			
Indiana	X		
Iowa	X		X
Kansas	X		X
Kentucky			
Louisiana	X		
Maine	X		
Maryland			
Massachusetts			
Michigan	X		
Minnesota	X		
Mississippi			
Missouri			
Montana		X	
Nebraska			X
Nevada			
New Hampshire	X		
New Jersey			
New Mexico	X		
New York			
North Carolina	X		
North Dakota	X		X
Ohio			
Oklahoma			
Oregon		X	
Pennsylvania			
Rhode Island	X		X

Table 1: Functions of State Directors

State/Territory	Combined EM/HS*	Separate EM & HS but Same Director	Adjutant General Directs EM and/or HS
South Carolina			X
South Dakota			
Tennessee			
Texas			
Utah	X		
Vermont			
Virginia			
Washington	X		X
West Virginia	X		
Wisconsin			X
Wyoming	X		
American Samoa			
District of Columbia	X		
Guam	X		
Puerto Rico			
U.S. Virgin Islands			X
*EM=Emergency Management; HS=Homeland Security.			

ALABAMA

Emergency Management

Director: Bruce Baughman

Functions/Responsibilities: Ala. Code § 31–9–4. The director, subject to the direction and control of the Governor, shall be the executive head of the Emergency Management Agency and shall be responsible to the Governor for carrying out the program for emergency management of this state. Ala. Code § 31–9–4. The director shall coordinate the activities of all organizations of emergency management within the state, and shall maintain liaison with and cooperate with major commanders of the armed forces within the state, the State Department of Public Safety, the State Military Department, and with emergency management agencies and organizations of other states and of the federal government, and shall have such additional authority, duties, and responsibilities authorized by this article as may be prescribed by the Governor. The director shall also hold the position of Assistant Director of Homeland Security for Emergency Preparedness and Response.

Qualifications/Statutory Authority: Ala. Code § 31–9–4. There is hereby created within the executive branch of the state government a department of emergency management, hereinafter called the "Emergency Management Agency," with a Director of Emergency Management, hereinafter called the "director," who shall be the head thereof. The director shall be appointed by the Governor.

Source: http://www.legislature.state.al.us/CodeofAlabama/1975/coatoc.htm (Title 31—Military Affairs and Civil Defense, chapter 9—Emergency Management).

Homeland Security

Director: James M. Walker, Jr.

Functions/Responsibilities: Ala. Code § 31–9A–5. The director, subject to the direction and authority of the Governor, shall be the executive head of the department and shall be responsible to the Governor for coordinating, designing, and implementing Alabama's program for homeland security. The director is given the following additional and cumulative powers and duties to: (1) Receive intelligence information from federal authorities relating to homeland security and ensure that, to the extent allowed by law, all appropriate and necessary intelligence and law enforcement information regarding homeland security is disseminated to and exchanged among appropriate executive departments responsible for homeland security, and where appropriate, promote the exchange of such information with county and local governments and private entities. (2) Assist in planning and executing exercises and simulations designed to practice those systems that would be utilized in response to a terrorist threat or attack within Alabama. (3) Assist in state efforts to ensure public health preparedness for a terrorist event. (4) Engage in the exchange of information with the federal government relating to immigration and efforts to improve the security of the borders, territorial waters, and ports of the United States.

Qualifications/Statutory Authority: Ala. Code § 31–9A–5. The position of Director of Homeland Security is created. The director shall be the head of the department. The director shall be appointed by and hold office at the pleasure of the Governor and shall be subject to confirmation by the Senate. Notwithstanding the foregoing, any person holding the position of director on June 18, 2003, shall not be subject to confirmation by the Senate.

Source: http://www.legislature.state.al.us/CodeofAlabama/1975/coatoc.htm (Title 31, chapter 9A—Alabama Homeland Security Act of 2003).

Adjutant General

Director: Major General Crayton M. Bowen

Functions/Responsibilities: Ala. Code § 31–2–58. The Adjutant General of the state shall be in direct charge of the military department and shall be responsible to the Governor and Commander in Chief for the proper performance of his duties. He shall supervise the receipt, preservation, repair, distribution, issuance and collection of all arms, military equipment and stores of the state and of the United States. He shall supervise all troops, arms and branches of the militia, such supervisory powers covering primarily all duties pertaining to organization, armament, discipline, training, recruiting, inspecting, instructing, pay, subsistence and supplies.

Qualifications: Ala. Code § 31–2–58. The Adjutant General shall be appointed from among active officers of the federally recognized National Guard, and he shall have had at least six years' service therein, two years of which must have been in the line, and he shall have served as a commissioned officer in the active National Guard for not less than four years.

Statutory Authority: Ala. Code § 31–2–58. The head of the Military Department shall be a commissioned officer of the National Guard of Alabama and shall be designated as the Adjutant General.

Source: http://www.legislature.state.al.us/CodeofAlabama/1975/31-2-58.htm (Title 31, chapter 2—Military Code).

ALASKA

Division of Homeland Security and Emergency Management

Director: John Madden

Functions/Responsibilities: Alaska Stat. § 26.20.025. The Alaska division of homeland security and emergency management, with the concurrence and approval of the adjutant general of the department, shall prepare and maintain a state homeland security plan and keep it current. The plan may include provisions for (1) investigation and assessment of attack threats to persons, facilities, systems, infrastructure, and other property in this state; (2) identification of geographical areas, municipalities, facilities, systems, infrastructure, or other property or persons especially vulnerable to an attack.

Qualifications: Appointed by governor. (Source: http://www.ak-prepared.com/DMVA/madden.htm.)

Statutory Authority: Alaska Stat. § 26.20.025. There is established in the department the Alaska division of homeland security and emergency management, possessing the powers and duties as set out in this section and as delegated by the adjutant general of the department.

Source: http://www.legis.state.ak.us/cgi-bin/folioisa.dll/stattx05/query=%5Bjump!3A!27as 2620025!27%5D/doc/%7B@11792%7D? (Title 26—Military Affairs and Veterans, chapter 20—Homeland Security and Civil Defense).

Adjutant General

Director: Major General Craig E. Campbell

Functions/Responsibilities: Alaska Stat. § 26.05.170. The governor's command is exercised through the adjutant general, who shall carry out the policies of the governor in military affairs. The adjutant general represents the governor and shall act in conformity with the governor's instructions. The adjutant general shall exercise control over the military department of the state. Alaska Stat. § 26.05.160. The adjutant general is the official liaison between the state and the active military in the state. The adjutant general shall provide advice and assistance to state agencies having dealings with the active military in the state. The adjutant general is the official liaison between the state and the federal Department of Veterans Affairs. The adjutant general shall provide advice and assistance to state agencies having dealings with the federal Department of Veterans Affairs.

Qualifications/Statutory Authority: Alaska Stat. § 26.05.160. The adjutant general of the state is appointed by the governor. The governor shall prescribe the grade of the adjutant general, which may not exceed major general. To be eligible for appointment as adjutant general, a person must be a citizen of the state.

Source: http://www.legis.state.ak.us/cgi-bin/folioisa.dll/stattx05/query=!22adjutant+general!22/ doc/%7B@11720%7D? (Title 26, chapter 5—Military Code of Alaska).

ARIZONA

Emergency Preparedness

Director: Louis Chaboya

Functions/Responsibilities: Ariz. Rev. Stat. § 26–305. The division shall prepare for and coordinate those emergency management activities that may be required to reduce the impact of disaster on persons or property. Ariz. Rev. Stat. § 26–306. The director shall, subject to the approval of the adjutant general: 1. Be the administrative head of the division. 2. Be the state director for emergency management. 3. Make rules necessary for the operation of the division. 4. Develop and test plans for meeting any condition constituting a state of emergency or state of war emergency, except those emergency plans specifically assigned by the governor to other state agencies.

Qualifications: Ariz. Rev. Stat. § 26–305. The adjutant general shall appoint the director who serves at the pleasure of the adjutant general. Selection of the director is on the basis of demonstrated ability in governmental functions or business administration and general knowledge of contingency planning and disaster preparedness.

Statutory Authority: Ariz. Rev. Stat. § 26–305. There is established in the department of emergency and military affairs the division of emergency management, which is administered by the department under the authority of the adjutant general, subject to powers vested in the governor as provided by law.

Source: http://www.azleg.state.az.us/ArizonaRevisedStatutes.asp?Title=26 (Title 26—Military Affairs and Emergency Management, chapter 2—Emergency Management).

Homeland Security

Director: Leesa Morrison

Functions/Responsibilities: Ariz. Rev. Stat. § 41–4252. The Arizona department of homeland security is established. The direction, operation and control of the department are the responsibility of the director. Ariz. Rev. Stat. § 41–4254. The department shall: Formulate policies, plans and programs to enhance the ability of this state to prevent and respond to acts of terrorism and other critical hazards. Develop a statewide homeland security strategy. Conduct preparedness training exercises to put state disaster plans into practice and identify shortcomings in the plans.

Qualifications/Statutory Authority: Ariz. Rev. Stat. § 41–4252–C. To be eligible for appointment as director a person must have a background or experience in one or more of the following areas: 1. Public administration. 2. Disaster response. 3. Law enforcement. 4. Business administration.

Source: http://www.azleg.state.az.us/ArizonaRevisedStatutes.asp?Title=41 (Title 41—State Government, chapter 41—Arizona Department of Homeland Security).

Adjutant General

Director: Major General David P. Rataczak

Functions/Responsibilities: Ariz. Rev. Stat. § 26–102–A. The adjutant general shall serve as head of the department. The adjutant general shall act as military chief of staff to the governor and chief of all branches of the militia. At the time of appointment, the adjutant general shall receive the state rank of major general and shall, at that time, become the ranking officer in the department of emergency and military affairs. Ariz. Rev. Stat. § 26–102–B. The adjutant general, as the military chief of staff, shall: 1. Act as military advisor to the governor and perform, as the governor prescribes, military duties not otherwise designated by law. 2. Adopt methods of administration for the national guard that are not inconsistent with laws and regulations of the United States department of defense or any subdivision of the United States department of defense. 3. Supervise and direct the organization, regulation, instruction and other activities of the national guard.

Qualifications: Ariz. Rev. Stat. § 26–101–D. The adjutant general shall be appointed by the governor pursuant to section 38–211 for a term of office of five years or to age sixty-four, whichever occurs first. The person appointed shall be a citizen of the United States and a resident of the state of Arizona. At the time of the appointment, the person appointed shall have qualifications required by the United States department of defense for the adjutant general and shall attain federal recognition in a grade not less than brigadier general within one year of the appointment. The adjutant general shall have served not less than five years in the national guard of Arizona in the last ten years. Failure to meet these qualifications or to retain federal recognition shall terminate the appointment.

Statutory Authority: Ariz. Rev. Stat. § 26–01. A. There is established a department of emergency and military affairs which shall consist of a division of emergency management and other divisions or offices as determined by the adjutant general pursuant to section 26–02, subsection C, paragraph 8. B. The department shall consist of the adjutant general and such other officers, warrant officers, enlisted personnel and employees as deemed necessary.

Source: http://www.azleg.state.az.us/ArizonaRevisedStatutes.asp?Title=26 (Title 26, Chapter 1—Emergency and Military Affairs).

ARKANSAS

Department of Emergency Management

Director: David Maxwell

Functions/Responsibilities: Ark. Code § 12–75–109. The Department of Emergency Management incorporates homeland security and preparedness functions. The Director of the Department of Emergency Management shall establish training and professional standards required to supplement state personnel based on state and federal disaster recovery program needs and shall establish a list of persons with those qualifications and make available to reserve cadre personnel such additional training and education opportunities as may be needed to maintain currency and proficiency in the needed skills. Based on the size, impact, and magnitude of the disaster event, the director shall determine the minimum number of reserve personnel required to effectively supplement regular state emergency management personnel and report these numbers to the Governor for approval.

Qualifications: Ark. Code § 12–75–109. The department shall have a director appointed by the Governor, with the advice and consent of the Senate, who shall serve at the pleasure of the Governor.

Statutory Authority: Ark. Code § 12–75–109. The Arkansas Department of Emergency Management is established as a public safety agency of the State of Arkansas.

Source: http://www.arkleg.state.ar.us/NXT/gateway.dll?f=templates&fn=default.htm&vid=blr: code (Title 12—Law Enforcement, Emergency Management, And Military Affairs, Subtitle 5— Emergency Management, chapter 75—Arkansas Emergency Services Act).

Adjutant General

Director: Major General Ronald S. Chastain

Functions/Responsibilities: Ark. Code § 12–61–106. In addition to being a state staff officer, the Adjutant General shall be the Chief-of-Staff to the Commander-in-Chief and the administrative head of the Military Department. He shall perform the duties prescribed for him in this code and in the regulations issued there under and in the statutes of the United States. He shall direct and supervise the functions and duties of the chief-of-staff departments. He shall hold office as provided in the National Defense Act as amended. He shall superintend the preparation of all returns and reports required by the United States from the state.

Qualifications: Ark. Code § 12–61–105. To be eligible for appointment, he shall be a citizen of the United States and a resident of the State of Arkansas, and: He must be an officer in the active militia with not less than seven (7) successive years' service immediately next preceding his appointment; or He must have been in service in the active militia of this state as a commissioned officer for a period of fifteen (15) years, eight (8) of which were as a field grade officer or general officer, or both combined; or He must have held the rank of a field grade officer in the active militia and, as such, have been called into federal service and have commanded a unit during such service.

Statutory Authority: Ark. Code § 12–61–105. There shall be an Adjutant General of the state who shall be appointed by the Governor and shall be a commissioned officer in the Adjutant General's department of the National Guard of this state and shall have rank not higher than major general.

Source: http://www.arkleg.state.ar.us/NXT/gateway.dll?f=templates&fn=default.htm&vid=blr: code (Title 12—Law Enforcement, Emergency Management, And Military Affairs, Subtitle 4— Military Affairs, chapter 61—Military Forces).

CALIFORNIA

Emergency Management

Director: Henry Renteria

Functions/Responsibilities: Cal. Gov't Code § 8585. The Director of the Office of Emergency Services, who shall also be the State Director of Civil Defense and the State Director of Emergency Planning, shall be in charge of the Office of Emergency Services and shall have all the rights and powers of a head of a department as provided by the Government Code. Cal. Gov't Code § 8587. During a state of war emergency, a state of emergency, or a local emergency, the director shall coordinate the emergency activities of all state agencies in connection with such emergency, and every state agency and officer shall cooperate with the director in rendering all possible assistance in carrying out the provisions of this chapter.

Qualifications: Cal. Gov't Code § 8585. The Director of the Office of Emergency Services shall be appointed by the Governor with the consent of the Senate, and shall serve at the pleasure of the Governor.

Statutory Authority: Cal. Gov't Code § 8585. There is in the office of the Governor the Office of Emergency Services, which office is the State Civil Defense Agency.

Source: http://law.justia.com/california/codes/gov/8585-8589.7.html (Title 2, chapter 7, article 5—Office of Emergency Services).

Homeland Security

Director: Matthew Bettenhausen

Functions/Responsibilities: Cal. Gov't Code § 12016. The Director of Homeland Security shall be in charge of homeland security and shall be the state coordinator of all homeland security activities, including, but not limited to, homeland security strategy, information analysis related to terrorism, and protection of critical infrastructure from terrorism. Cal. Gov't Code § 9147.5. (a) Notwithstanding Section 7550.5, the Director of Homeland Security, in collaboration with the State Department of Health Services, shall, on or before February 1 of each year, report to the chairperson of the Joint Legislative Budget Committee, the chairperson of the transportation committee of each house of the Legislature, and the chairperson of the budget committee of each house of the Legislature, on their respective expenditures of federal homeland security and bioterrorism funds.

Qualifications/Statutory Authority: Cal. Gov't Code § 12016. The Governor shall appoint, to serve at his or her pleasure, an executive officer who shall be Director of Homeland Security.

Source: http://www.leginfo.ca.gov/calaw.html (Division 3—Executive Department, Part 2—Constitutional Officers).

Adjutant General

Director: Major General William H. Wade II

Functions/Responsibilities: Cal. Mil. & Vet. Code § 160. The Adjutant General is chief of staff to the Governor, subordinate only to the Governor and is the commander of all state military forces. Cal. Mil. & Vet. Code § 171. The Adjutant General shall keep a register of all the officers of the militia of the State and shall keep in his office all records and papers required to be kept and filed therein. § 172. The Adjutant General shall make a report to the Governor every fourth year, commencing in 1963, the report to include a statement of the moneys received and disbursed by the Adjutant General for military purposes, the number and condition of the active militia, and a history of the activities and developments of the Military Department during the preceding four years.

Qualifications: Cal. Mil. & Vet. Code § 162. The Adjutant General shall be appointed by the Governor with the advice and consent of the Senate, and shall hold office at the pleasure of the Governor, or until his successor is appointed and has qualified. No person is eligible for appointment as Adjutant General unless he had not less than a total of ten (10) years of commissioned service in the National Guard of the United States, of which at least four (4) years shall be service as a field grade officer in the California National Guard within the preceding 10-year period prior to the date of appointment and of which at least four (4) years shall have been in command of army or air troops at the battalion or equivalent or higher command level or four (4) years as a staff officer at brigade or equivalent or higher staff level.

Statutory Authority: Cal. Mil. & Vet. Code § 160.5. Any statute of this state referring to the Commanding General of the State Military Forces is deemed to refer to the Adjutant General.

Source: http://www.leginfo.ca.gov/cgi-bin/displaycode?section=mvc&group=00001-01000&file=160-190 (Division 2—The Military Forces of the State, chapter 2, article 3—The Adjutant General).

COLORADO

Emergency Management

Director: David Holm

Functions/Responsibilities: Colo. Rev. Stat. § 24–32–2102. Authorize and provide for coordination of activities relating to disaster prevention, preparedness, response, and recovery by agencies and officers of this state and similar state-local, interstate, federal-state, and foreign activities in which the state and its political subdivisions may participate. Colo. Rev. Stat. § 24–32–2105. (1) The division of emergency management and the office of the director shall exercise their powers and perform their duties and functions under the department of local affairs and the executive director as if the same were transferred to the department by a type 2 transfer, as such transfer is defined in the "Administrative Organization Act of 1968", article 1 of this title. (2) The division shall prepare and maintain a state disaster plan which complies with all applicable federal and state regulations and shall keep such plan current.

Qualifications/Statutory Authority: Colo. Rev. Stat. § 24–32–2105. There is hereby created in the department of local affairs the division of emergency management, referred to in this part 21 as the "division". Pursuant to section 13 of article XII of the state constitution, the executive director shall appoint a director, referred to in this part 21 as the "director", as head of the division.

Source: http://198.187.128.12/colorado/lpext.dll?f=templates&fn=fs-main.htm&2.0 (Title 24—Government-State, article 32—Department of Local Affairs, Part 21—Office of Disaster Emergency Services).

Homeland Security

Director: David Holm

Functions/Responsibilities: Colo. Rev. Stat. § 24–33.5–1605. (1) (a) The creation and implementation of the terrorism preparedness plan described in section 24–33.5–1604; and (b) The prevention and detection of terrorist training activities described in section 18–9–120, C.R.S. Colo. Rev. Stat. § 24–33.5–1604. (1) (a) To inquire into the threat of terrorism in Colorado and the state of preparedness to respond to that threat and to make recommendations to the governor and the general assembly; (b) To cooperate with the federal office of homeland security and other agencies of the federal government and other states in matters related to terrorism.

Qualifications: Colo. Rev. Stat. § 24–33.5–1603. The director of the office of preparedness, security, and fire safety, referred to in this part 16 as the "director," shall be appointed by the executive director pursuant to section 13 of article XII of the state constitution.

Statutory Authority: Colo. Rev. Stat. § 24–33.5–1601. An agency should be established in the state government to coordinate Colorado's response to the threat of terrorism.

Source: http://www.michie.com/colorado/lpext.dll?f=templates&fn=main-h.htm&cp=cocode (Title 24, article 33.5, Part 16—Office of Preparedness, Security and Fire Safety).

Adjutant General

Director: Major General Mike Edwards

Functions/Responsibilities: Colo. Rev. Stat. § 28–3–106. (1) (a) The adjutant general shall be the chief of staff to the commander in chief and the administrative head of the department of military and veterans affairs. (b) He or she shall have custody of all military records, correspondence, and other military documents. He or she shall be the medium of military correspondence with the governor and perform all other duties pertaining to his or her office prescribed by law. (c) The adjutant general shall prepare and transmit annually, in the form and manner prescribed by the heads of the principal departments pursuant to the provisions of section 24–1–136, C.R.S., a report accounting to the governor and the state, veterans, and military affairs committees of the house of representatives and the senate for the efficient discharge of all responsibilities assigned by law or directive to the adjutant general.

Qualifications/Statutory Authority: Colo. Rev. Stat. § 28–3–105. There shall be an adjutant general of the state who shall be appointed by the governor, with the advice and consent of the senate, who shall be a staff officer, who at the time of appointment shall be a commissioned officer of the National Guard of this state with not fewer than ten years' military service in the armed forces of this state or of the United States at least five of which have been commissioned service in the Colorado National Guard, and who has attained the grade of lieutenant colonel or a higher grade with federal recognition in such grade at least one year prior to his or her appointment as adjutant general. The adjutant general shall serve at the pleasure of the governor.

Source: http://198.187.128.12/colorado/lpext.dll/Infobase4/1/496b5/496dc/496de?f=templates& fn=fs-main-doc.htm&q=%22adjutant%20general%22&x=Advanced&2.0#LPHit1 (Title 28— Military and Veterans, article 3—National Guard).

CONNECTICUT

Department of Emergency Management and Homeland Security

Director: James M. Thomas

Functions/Responsibilities: Conn. Gen. Stat. § 28–1a. The commissioner shall be responsible for: (1) Coordinating with state and local government personnel, agencies and authorities and the private sector to ensure adequate planning, equipment, training and exercise activities by such personnel, agencies and authorities and the private sector with regard to homeland security; (2) coordinating, and as may be necessary, consolidating homeland security communications and communications systems of the state government with state and local government personnel, agencies and authorities, the general public and the private sector; (3) distributing and, as may be appropriate, coordinating the distribution of information and security warnings to state and local government personnel, agencies and authorities and the general public.

Qualifications/Statutory Authority: Conn. Gen. Stat. § 28–1a. There is established a Department of Emergency Management and Homeland Security. Said department shall be the designated emergency management and homeland security agency for the state. The department head shall be the commissioner, who shall be appointed by the Governor in accordance with the provisions of sections 4–5, 4–6, 4–7 and 4–8 with the powers and duties prescribed in said sections. The commissioner shall possess professional training and knowledge consisting of not less than five years of managerial or strategic planning experience in matters relating to public safety, security, emergency services and emergency response. No person possessing a record of any criminal, unlawful or unethical conduct shall be eligible for or hold such position. Any person with any present or past political activities or financial interests that may substantially conflict with the duties of the commissioner or expose such person to potential undue influence or compromise such person's ability to be entrusted with necessary state or federal security clearances or information shall be deemed unqualified for such position and shall not be eligible to hold such position.

Source: http://search.cga.state.ct.us/dtsearch_pub_statutes.html (Title 28—Civil Preparedness and Emergency Services, chapter 517—Civil Preparedness, Department of Emergency Management and Homeland Security).

Adjutant General

Director: Major General Thaddeus Martin

Functions/Responsibilities: Conn. Gen. Stat. § 27–19. The Military Department shall be under the charge of the Adjutant General. Conn. Gen. Stat. § 27–20. The Adjutant General shall make such returns and reports to such officers as may be prescribed by the Department of Defense in regulations pertaining to the National Guard and naval militia, at such times and in such form as may, from time to time, be prescribed. The Adjutant General shall keep a record of all officers and enlisted personnel and shall also keep in the office all records and papers required by law or regulations to be filed therein.

Qualifications: Conn. Gen. Stat. § 27–19. The Adjutant General shall have had at least ten years' commissioned service in the armed forces of the United States. No person shall be appointed or continue to serve after reaching the age of sixty-four years.

Statutory Authority: Conn. Gen. Stat. § 27–19. On or before July 1, 1980, the Governor shall appoint an Adjutant General with the rank of major general to serve for a term of two years from July 1, 1980.

Source: http://search.cga.state.ct.us/dtsearch_pub_statutes.html (Title 27—Armed Forces and Veterans, chapter 504—Militia).

DELAWARE

Department of Emergency Management (DEMA)

Director: James E. Turner III

Functions/Responsibilities: Del. Code Title 20, § 3106. (1) The Director, as head of the agency, shall supervise, on a full-time basis, all fiscal, planning, administrative, operational and other functions of DEMA as assigned by law or the Secretary. The Director, subject to the direction and control of the Governor, shall be the executive head of DEMA and shall be responsible to the Secretary of Public Safety for carrying out the program for emergency management of this State. (2) Prepare and maintain a comprehensive plan and program for the emergency management of the State, such plan to be integrated into and coordinated with the emergency management plans of the federal government and of other states and political subdivisions of this State to the fullest possible extent.

Qualifications/Statutory Authority: Del. Code Title 20, § 3105. To assure the prompt, proper and effective discharge of basic state responsibilities relating to emergency management, there is hereby formally created the Delaware Emergency Management Agency of the Department of Safety and Homeland Security (hereinafter also referred to as "DEMA"). § 3106. The Delaware Emergency Management Agency of the Department of Safety and Homeland Security, within the executive branch of the state government, shall consist of and be organized substantially as follows: Director and Deputy Principal Assistant.—The Director and the Deputy Principal Assistant ("Deputy") shall be appointed by the Secretary of Public Safety with the written approval of the Governor.

Source: http://delcode.delaware.gov/title20/c031/sc02/index.shtml (Title 20—Military and Civil Defense, chapter 31—Emergency Management, Subchapter II - DEMA).

Department of Safety and Homeland Security

Director: David Mitchell

Functions/Responsibilities: Del. Code Title 29, § 8202. The administrator and head of the Department shall be the Secretary of the Department of Safety and Homeland Security, who shall be a person qualified by training and experience to perform the duties of the office and preference shall be given to a resident of this State provided that such person is acceptable and equally qualified. Del. Code Title 29, § 8203. The Secretary may: Supervise, direct and account for the administration and operation of the Department, its divisions, subdivisions, offices, functions and employees. Appoint such additional personnel as may be necessary for the administration and operation of the Department within such limitations as may be imposed by law.

Qualifications: Del. Code Title 29, § 8202. In the event the position of Secretary is vacant, the Governor, by appointment, shall have the power to fill the position or positions of division director as are vacant. Directors so appointed shall serve at the pleasure of the Governor and, upon the position of Secretary being filled, such directors may be removed by the Secretary with the written approval of the Governor.

Statutory Authority: Del. Code Title 29, § 8201. A Department of Safety and Homeland Security is established. Title 29 § 8202. The administrator and head of the Department shall be the Secretary of the Department of Safety and Homeland Security.

Source: http://delcode.delaware.gov/title29/c082/index.shtml (Title 29—State Government, chapter 82—Department of Safety and Homeland Security).

Adjutant General

Director: Major General Francis D. Vavala

Functions/Responsibilities: Del. Code Title 20, § 122. The Adjutant General shall be responsible to the Commander in Chief for carrying out the policies of the Commander in Chief and shall issue orders in the Commander in Chief's name. Del. Code Title 20, § 123. (1)

Supervise all troops, arms and branches of the Delaware National Guard, Delaware State Defense Forces and militia, with such supervisory powers as are necessary, covering primarily all duties pertaining to their organization, armament, discipline, training, recruiting, inspection, pay, subsistence and supplies. (2) Supervise the receipt, preservation, repair, distribution, issue and collection of all arms and military stores of this State.

Qualifications: Del. Code Title 20, § 122. To be eligible for appointment to the office of Adjutant General a person must have served as a commissioned officer of the Delaware National Guard or the armed forces of the United States.

Statutory Authority: Del. Code Title 20, § 121. The Delaware National Guard, the Delaware State Defense Forces and the Delaware militia, when not in the service of the United States, shall be governed and their affairs administered pursuant to the laws of this State, and the laws of the United States, by the Governor as Commander in Chief, through the Department of Military Affairs, which is hereby established and which shall consist of the Adjutant General as its chief executive and such other officers, warrant officers and enlisted personnel and civilian employees as the laws of the State or the laws of the United States may direct or permit.

Source: http://delcode.delaware.gov/title29/c082/index.shtml (Title 20, chapter 1—Delaware National Guard, subchapter II—Department of Military Affairs).

FLORIDA

Emergency Management

Director: W. Craig Fugate

Functions/Responsibilities: Fla. Stat. Title XVII § 252.32. To provide the means to assist in the prevention or mitigation of emergencies which may be caused or aggravated by inadequate planning for, and regulation of, public and private facilities and land use. Fla. Stat. Title XVII § 252.35. (1) The division is responsible for maintaining a comprehensive statewide program of emergency management. The division is responsible for coordination with efforts of the Federal Government with other departments and agencies of state government, with county and municipal governments and school boards, and with private agencies that have a role in emergency management. (2) The division is responsible for carrying out the provisions of ss. 252.31–252.90. In performing its duties under ss. 252.31–252.90, the division shall: Prepare a state comprehensive emergency management plan, which shall be integrated into and coordinated with the emergency management plans and programs of the Federal Government.

Qualifications: Appointed by governor. (Source: http://www.floridadisaster.or/director_office/ bio_files/craig_bio.htm.)

Statutory Authority: Fla. Stat. Title XVII § 252.32. To create a state emergency management agency to be known as the "Division of Emergency Management," to authorize the creation of local organizations for emergency management in the political subdivisions of the state, and to authorize cooperation with the Federal Government and the governments of other states.

Source: http://www.leg.state.fl.us/statutes/index.cfm?Mode=ViewStatutes&Submenu=1 (Title XVII—Military Affairs, chapter 252—Emergency Management).

Department of Law Enforcement

Director: Gerald M. Bailey

Functions/Responsibilities: Fla. Stat. Title XLVII § 943.0311. (1) (a) Coordinate the efforts of the department in the ongoing assessment of this state's vulnerability to, and ability to detect, prevent, prepare for, respond to, and recover from acts of terrorism within or affecting this state. (b) Prepare recommendations for the Governor, the President of the Senate, and the Speaker of the House of Representatives, which are based upon ongoing assessments to limit the vulnerability of the state to terrorism. (c) Coordinate the collection of proposals to limit the vulnerability of the state to terrorism. (2) The chief shall conduct or cause to be conducted by the personnel and with the resources of the state agency, state university, or community college that owns or leases a building, facility, or structure, security assessments of buildings, facilities, and structures owned or leased by state agencies, state universities, and community colleges using methods and instruments made available by the department.

Qualifications: Fla. Stat. Title XLVII § 943.03. The executive director shall have served at least 5 years as a police executive or possess training and experience in police affairs or public administration and shall be a bona fide resident of the state.

Statutory Authority: Fla. Stat. Title XLVII § 943.03101. The Legislature finds that with respect to counter-terrorism efforts and initial responses to acts of terrorism within or affecting this state, specialized efforts of emergency management that are unique to such situations are required and that these efforts intrinsically involve very close coordination of federal, state, and local law enforcement agencies with the efforts of all others involved in emergency-response efforts. In order to best provide this specialized effort with respect to counter-terrorism efforts and responses, the Legislature has determined that such efforts should be coordinated by and through the Department of Law Enforcement, working closely with the Division of Emergency Management and others involved in preparation against acts of terrorism in or affecting this state, and in the initial response to such acts, in accordance with the state comprehensive emergency management plan prepared pursuant to Title XVIII § 252.35(2)(a).

Source: http://law.justia.com/florida/codes/TitleXLVII/ch0943.html (Title XLVII—Criminal Procedure and Corrections, chapter 943—Department of Law Enforcement).

Adjutant General

Director: Major General Douglas Burnett

Functions/Responsibilities: Fla. Stat. Title XVII, § 250.05. The head of the Department of Military Affairs is the Adjutant General. Fla. Stat. Title XVII, § 250.10. The responsibilities of the Adjutant General are to: (a) supervise the receipt, preservation, repair, distribution, issue, and collection of all arms and military equipment of the state; (b) supervise all troops and branches of the Florida National Guard, including their organization, armament, discipline, training, recruiting, inspection, instruction, pay, subsistence, and supplies.

Qualifications: Fla. Stat. Title XVII, § 250.10. In case of a vacancy, the Governor shall, subject to confirmation by the Senate, appoint a federally recognized officer of the Florida National Guard, who has served in the Florida National Guard for the preceding 5 years and attained the rank of colonel or higher, to be the Adjutant General of the state with the rank of not less than brigadier general or such higher rank as authorized by applicable tables of organization of the Department of the Army or the Department of the Air Force.

Statutory Authority: Fla. Stat. Title XVII, § 250.05. The head of the Department of Military Affairs is the Adjutant General.

Source: http://www.flsenate.gov/Statutes/index.cfm?App_mode=Display_Statute&URL=Ch02 50/ch0250.htm (Title XVII, chapter 250—Military Affairs).

GEORGIA

Emergency Management

Director: Charley English

Functions/Responsibilities: Ga. Code § 38–3–20. The director, subject to the direction and control of the Governor, shall be the executive head of the Georgia Emergency Management Agency and shall be responsible to the Governor for carrying out the program for emergency management in this state. He or she shall coordinate the activities of all organizations for emergency management within the state, shall maintain liaison with and cooperate with emergency management agencies and organizations of other states and of the federal government, and shall have such additional authority, duties, and responsibilities.

Qualifications: Ga. Code § 38–3–20. The Governor shall appoint the director of emergency management. He or she shall hold office at the pleasure of the Governor, who shall fix his or her compensation. The director of emergency management shall hold no other state office.

Statutory Authority: Ga. Code § 38–3–20. There is established the Georgia Emergency Management Agency with a director of emergency management who shall be the head thereof. The Georgia Emergency Management Agency shall be assigned to the Office of Planning and Budget for administrative purposes only as provided in Code Section 50–4–3.

Source: http://www.lexis-nexis.com/hottopics/gacode/default.asp. (Title 38—Military, Emergency Management, and Veterans Affairs, chapter 3—Emergency Management).

Office of Homeland Security

Director: Charley English

Functions/Responsibilities: Exec. Order 8–25–04–01 (August 2004). Safeguarding the citizens of this state against the possibility of a threat or act of terrorism. (Source: http://www.gov.state. ga.us/ExOrders/08_25_04_01.pdf.) Homeland Security Mission statement: The mission of the Georgia Office of Homeland Security is to lead and direct the preparation, employment and management of state resources to safeguard Georgia and its citizens against threats or acts of terrorism and the effects of natural disasters. (Source: http://www.gema.state.ga.us/ohsgemaweb. nsf/9C891F3A609DCA46852570C8005A2D64/EB5D09226CEA56E085257114006DA750?Op enDocument.)

Qualifications: Appointed by governor. (Source: http://www.gema.state.ga.us/ohsgemaweb.nsf/ 72b97e7aee4095b98525711b00581aea/f786de8a5bae92c1852573640063f621?OpenDocument.)

Statutory Authority: Exec. Order 8–25–04–01 (August 2004). That the Executive Commander of the Homeland Security Central Command shall be the Governor's Director of Homeland Security.

Source: http://www.gov.state.ga.us/ExOrders/08_25_04_01.pdf.

Adjutant General

Director: Major General Terry Nesbitt

Functions/Responsibilities: Ga. Code § 38–2–151. (a) The adjutant general shall be chief of staff to the Governor and subordinate only to the Governor in matters pertaining to the Department of Defense and the military and naval affairs of the state. (b) Whenever the Governor and those who would act in succession to him under the Constitution and laws of the state are unable to perform the duties of commander in chief, the adjutant general shall command the militia. (c) It shall be the duty of the adjutant general: (1) To direct the planning and employment of the forces of the organized militia in carrying out their state military mission; (2) To establish unified command of state forces whenever they are jointly engaged; and (3) To coordinate the military and naval affairs with the emergency management agency of the state.

Qualifications/Statutory Authority: Ga. Code § 38–2–150. There shall be an adjutant general of the state who shall be appointed by the Governor for a term concurrent with the term of the Governor appointing such person and who shall serve as such at the pleasure of the Governor. The adjutant general shall have not less than the rank of a major general, the specific rank to be determined by the Governor. The adjutant general shall not be less than 30 nor more than 65 years of age. No person shall be eligible to hold the office of adjutant general unless he or she holds or has held a commission of at least the rank of field grade or the equivalent in the organized militia of the state, in the armed forces of the United States, or in a reserve component thereof and shall have served not less than five years in one or more of such services at the time of his or her appointment.

Source: http://www.lexis-nexis.com/hottopics/gacode/default.asp. (Title 38, chapter 2—Military Affairs).

HAWAII

State Civil Defense

Director: Major General Robert G. F. Lee

Functions/Responsibilities: Haw. Rev. Stat. § 128–3. (a) The director of civil defense may, from funds allotted therefor, employ technical, clerical, stenographic, and other personnel and make such expenditures as may be necessary. (b) The director, subject to the direction and control of the governor, shall be the executive head of the civil defense agency. The director shall coordinate the activities of all organizations for civil defense within the State, public or private, and shall maintain liaison with and cooperate with other civil defense agencies as provided in this chapter. (d) The director shall, with the approval of the county council, appoint for each political subdivision a deputy director who may be removed by the director.

Qualifications: Article V State Constitution. Appointed by the governor.

Statutory Authority: Haw. Rev. Stat. § 128–1. Because of the importance of the State as a strategic defense area, the dependence of the State upon seaborne commerce for food supplies and other commodities essential to the public health, safety, and welfare and to the economic life of its people, the danger of shortages of such supplies and commodities, and other emergency conditions affecting the readiness of this community to do its part in the existing national emergency which was declared by the President on December 16, 1950, and the possibility of disasters or emergencies of great destructiveness resulting from enemy attack, sabotage, or other hostile action, therefore in order to insure that preparations of this State and the government provided for this State will be adequate to deal with disasters or emergencies, to make adequate provision against such shortages, to maintain the strength, resources, and economic life of the community and provide for prompt and effective action, as the circumstances develop and in cooperation with the federal government, to further and promote the national defense and civil defense and to protect the public health, safety, and welfare, the provisions of this chapter are hereby found and declared to be necessary.

Source: http://www.capitol.hawaii.gov/site1/hrs/default.asp (Chapter 128, Civil Defense and Emergency Act).

Adjutant General

Director: Major General Robert G. F. Lee

Functions/Responsibilities: Haw. Rev. Stat. Title 10 § 121–9. The adjutant general shall perform such duties as are prescribed by law and such other military duties consistent with the regulations and customs of the armed forces of the United States as required by the governor. The adjutant general shall supervise all of the forces comprising the military components of the department of defense of the State. The supervisory power shall include the command, discipline, training, and recruiting of the armed forces of the State, military operations, distribution of troops, inspections, armament, military education and instruction, fiscal operations, administration, and supply. The adjutant general is authorized to confer the powers of police officers, including the power to arrest, to employees of the department who are engaged as security guards for national guard and civil defense facilities; provided that such powers shall remain in force and effect only while the security guards are in the actual performance of their duties as security guards.

Qualifications/Statutory Authority: Haw. Rev. Stat. title 10 § 121–7. The adjutant general shall be appointed and be subject to removal as set forth in section 26–31. The adjutant general shall serve for the term as set forth in section 6, article V, of the Constitution. No person shall be eligible for appointment as adjutant general unless the person holds or has held a commission of at least a field grade officer, federally recognized as such, or its equivalent in the national guard, state defense force, or other branch of the armed forces of this or any other state or territory of the United States, or in the armed forces of the United States or a reserve component thereof and has served as a commissioned officer in one or more of the armed services for at least ten years.

Source: http://www.capitol.hawaii.gov/hrscurrent/Vol03_Ch0121-0200D/HRS0121/HRS_0121-.HTM (Title 10—Public Safety and Internal Security, chapter 121—Militia; National Guard).

IDAHO

Bureau of Homeland Security and Disaster Emergencies

Director: William Bishop

Functions/Responsibilities: Idaho Code § 46–1006. (1) In all matters of disaster services, the adjutant general shall represent the governor and shall on behalf of the governor, coordinate the activities of all of the state agencies in disaster services. (2) The bureau shall prepare, maintain and update a state disaster plan based on the principle of self-help at each level of government. The plan may provide for: (a) Prevention and minimization of injury and damage caused by disaster; (b) Prompt and effective response to disaster; (c) Emergency relief; (d) Identification of areas particularly vulnerable to disasters; (e) Assistance to local officials in designing local emergency action plans; (f) Authorization and procedures for the erection or other construction of temporary works designed to protect against or mitigate danger, damage, or loss from disaster; (g) Preparation and distribution to the appropriate state and local officials of catalogs of federal, state and private assistance programs; (h) Assistance to local officials in designing plans for search, rescue, and recovery of persons lost, entrapped, victimized, or threatened by disaster; (i) Organization of manpower and chains of command; (j) Coordination of federal, state, and local disaster activities; (k) Coordination of the state disaster plan with the disaster plans of the federal government.

Qualifications: Idaho Code § 46–1005. The bureau may be headed by a chief appointed by the adjutant general with the concurrence of the governor or the governor may appoint the adjutant general to serve as chief.

Statutory Authority: Idaho Code § 46–1004. Within the military division of the office of governor, a bureau of homeland security is established.

Source: http://www3.state.id.us/idstat/TOC/46010KTOC.html (Title 46—Militia and Military Affairs, chapter 10—State Disaster Preparedness Act).

Adjutant General

Director: Major General Lawrence F. Lafrenz

Functions/Responsibilities: Exec. Order No. 2003–08 (August 2003). The Adjutant General of the State of Idaho is the Governor's authorized representative for emergency planning, preparedness, response, and recovery from all hazards including paramilitary acts, such as terrorism and the use of weapons of mass destruction. (Source: http://gov.idaho.gov/mediacenter/execorders/eo03/eo_2003_08.htm.) Idaho Code § 46-112. To be chief of staff to the commander-in-chief and administrative head of the military division of the office of governor. To be custodian of all military records and property of the national guard and organized militia. To publish and distribute all orders from the governor as commander-in-chief and perform such other duties as the governor may direct. To supervise the training of the national guard and the organized militia.

Qualifications/Statutory Authority: Idaho Code § 46–111. There shall be an adjutant general who shall be appointed by the governor and shall hold office during the pleasure of the governor and his commission shall expire with the term of the governor appointing him. The adjutant general shall be commissioned in the national guard with the rank of not less than brigadier general. No person is eligible for appointment as adjutant general unless he is a federally recognized member of the national guard with current service of not less than six (6) years as a commissioned officer in the national guard of Idaho and has attained the rank of colonel or above.

Source: http://www3.state.id.us/idstat/TOC/46001KTOC.html (Title 46, chapter 1—State Militia).

ILLINOIS

Emergency Management

Director: Andrew Velasquez III

Functions/Responsibilities: Ill. Comp. Stat. 20 § 3305/5. The Director, subject to the direction and control of the Governor, shall be the executive head of the Illinois Emergency Management Agency and the State Emergency Response Commission and shall be responsible under the direction of the Governor, for carrying out the program for emergency management of this State. The Director shall also maintain liaison and cooperate with the emergency management organizations of this State and other states and of the federal government.

Qualifications/Statutory Authority: Ill. Comp. Stat. 20 § 3305/5. There is created within the executive branch of the State Government an Illinois Emergency Management Agency and a Director of the Illinois Emergency Management Agency, herein called the "Director" who shall be the head thereof. The Director shall be appointed by the Governor, with the advice and consent of the Senate.

Source: http://www.ilga.gov/legislation/ilcs/ilcs3.asp?ActID=368&ChapAct=20%26nbsp%3BILCS%26nbsp%3B3305%2F&ChapterID=5&ChapterName=EXECUTIVE+BRANCH&ActName=Illinois+Emergency+Management+Agency+Act%2E. (Title 20—Executive Branch, section 3305—Emergency Management Agency).

Terrorism Task Force

Director: Jill Morgenthaler

Functions/Responsibilities: ITTF mission statement. The mission of the Illinois Terrorism Task Force is to implement a comprehensive coordinated strategy for domestic preparedness in the state of Illinois, bringing together agencies, organizations and associations representing all disciplines in the war against terrorism. (Source: http://www.illinoishomelandsecurity.net/ittf/ittfmission.htm.) Exec. Order 17 (2003). II. A. The Illinois Terrorism Task Force, as an advisory body to the Governor and the Deputy Chief of Staff for Public Safety, shall develop and recommend to the Governor the State's domestic terrorism preparedness strategy. B. The Illinois Terrorism Task Force shall develop policies related to the appropriate training of local, regional and State officials to respond to terrorist incidents involving conventional, chemical, biological and/or nuclear weapons. C. The Illinois Terrorism Task Force shall oversee the weapons of mass destruction teams, which the Governor may deploy in the event of a terrorist attack to assist local responders and to coordinate the provision of additional State resources.

Qualifications: Exec. Order 17 (2003). I. D. The Governor shall appoint a Chair to serve as the administrator of the Illinois Terrorism Task Force. The Chair shall report to the Deputy Chief of Staff for Public Safety on all activities of the Illinois Terrorism Task Force. The Chair shall also serve as a policy advisor to the Deputy Chief of Staff for Public Safety on matters related to Homeland Security.

Statutory Authority: Exec. Order 17 (2003). I. A. I hereby establish the Illinois Terrorism Taskforce as an advisory body, reporting directly to the Governor and to the Deputy Chief of Staff for Public Safety.

Source: http://www.il.gov/gov/execorder.cfm?eorder=17.

Adjutant General

Director: Major General Randal E. Thomas

Functions/Responsibilities: Ill. Comp. Stat. 20 § 1805/22. The Adjutant General shall be charged with carrying out the policies of the Commander-in-Chief and shall issue orders in his name. Orders of The Adjutant General shall be considered as emanating from the Commander-in-Chief. He shall be the immediate adviser of the Commander-in-Chief on all matters relating to the militia and shall be charged with the planning, development and execution of the program of the military forces of the State. Ill. Comp. Stat. 20 § 1805/22–1. The Adjutant General has the power and authority to enter into contracts and agreements in the name of the State of Illinois with the Federal government on any and all matters relating to the organizing, training, equipping, quartering and maintenance of the Illinois National Guard. The Adjutant General shall keep a record of the appointments of officers, warrant officers, and noncommissioned officers.

Qualifications: Ill. Comp. Stat. 20 § 1805/16. The Adjutant General and the Assistant Adjutants General shall have had 10 or more years of active commissioned service in a component of the U.S. Armed Forces, the active Illinois Army National Guard, or active Illinois Air National Guard, as appropriate, and have attained at least the grade of or equivalent to Colonel or Lieutenant Colonel, respectively.

Statutory Authority: Ill. Comp. Stat. 20 § 1805/20. There is hereby established in the Executive Branch of the State Government, a principal department which shall be known as the Department of Military Affairs. The Department of Military Affairs shall consist of The Adjutant General, Chief of Staff; an Assistant Adjutant General for Army; an Assistant Adjutant General for Air; and the number of military and civilian employees required.

Source: http://www.ilga.gov/legislation/ilcs/ilcs4.asp?DocName=002018050HArt%2E+IV&ActID=315&ChapAct=20%26nbsp%3BILCS%26nbsp%3B1805%2F&ChapterID=5&ChapterName=EXECUTIVE+BRANCH&SectionID=37474&SeqStart=5900&SeqEnd=7800&ActName=Military+Code+of+Illinois%2E. (Title 20, Section 1805—Military Code of Illinois).

INDIANA

Department of Homeland Security

Director: J. Eric Dietz

Functions/Responsibilities: In 2005 the State of Indiana consolidated all of its emergency management and homeland security efforts into one department by creating the Indiana Department of Homeland Security (IDHS). Ind. Code § 10–19–3–3. Serve as the chief executive and administrative officer of the department. Serve as the state coordinating officer under federal law for all matters relating to emergency and disaster mitigation, preparedness, response, and recovery. Use and allocate the services, facilities, equipment, personnel, and resources of any state agency, on the governor's behalf, as is reasonably necessary in the preparation for, response to, or recovery from an emergency or disaster situation that threatens or has occurred in Indiana. Develop a plan to protect key state assets and public infrastructure from a disaster or terrorist attack.

Qualifications: Ind. Code § 10–19–3–1. The governor shall appoint an individual to be the executive director of the department.

Statutory Authority: Ind. Code § 10–19–2–1. The department of homeland security is established. IC 10–19–2–2. The department consists of the following divisions: (1) The division of planning and assessment. (2) The division of preparedness and training. (3) The division of emergency response and recovery. (4) The division of fire and building safety.

Sources: http://www.in.gov/legislative/ic/code/title10/ar19/ (Title 10—Public Safety, article 19—Department of Homeland Security); Department Web site, http://www.in.gov/dhs/3979.htm.

Adjutant General

Director: Major General R. Martin Umbarger

Functions/Responsibilities: Ind. Code § 10–16–2–8. The adjutant general shall do the following: Execute all orders given by the commander in chief. § 10–16–2–5. The adjutant general shall be ex officio chief of staff. Ind. Code § 10–16–2–9. The adjutant general shall superintend the preparation of all returns and reports required by the United States from the state. The adjutant general shall attend to the safekeeping and repairing of the ordnance, arms, accouterments, equipment, and all other military and naval property belonging to the state or issued to it by the United States.

Qualifications: Ind. Code § 10–16–2–6. (a) The governor shall appoint the adjutant general. (b) The adjutant general must hold the rank of not less than brigadier general. (c) The governor may increase the rank of the adjutant general not to exceed the rank of major general as a reward for efficient and loyal service to the state.

Statutory Authority: Ind. Code § 10–16–2–1. (a) The military department of the state: is established. (b) The military department consists of the following: An adjutant general, who shall be the executive and administrative head of the department.

Source: http://www.in.gov/legislative/ic/code/title10/ar16/ch2.html. (Title 10—Public Safety, article 16—Indiana Military Code).

IOWA

Homeland Security & Emergency Management Division

Director: Major General G. Ron Dardis

Functions/Responsibilities: Iowa Code 1 § 29C.7. The adjutant general, as the director of the department of public defense and under the direction and control of the governor, shall have supervisory direction and control of the homeland security and emergency management division and shall be responsible to the governor for the carrying out of the provisions of this chapter. Iowa Code 1 § 29C.8. The administrator, upon the direction of the governor and supervisory control of the director of the department of public defense, shall: a. Prepare a comprehensive plan and emergency management program for homeland security, disaster preparedness, response, recovery, mitigation, emergency operation, and emergency resource management of this state. b. Make such studies and surveys of the industries, resources, and facilities in this state as may be necessary to ascertain the vulnerabilities of critical state infrastructure and assets to attack and the capabilities of the state for disaster recovery, disaster planning and operations, and emergency resource management, and to plan for the most efficient emergency use thereof.

Qualifications: Iowa Code 1 § 29A.11. There shall be an adjutant general of the state who shall be appointed and commissioned by the governor subject to confirmation by the senate and who shall serve at the pleasure of the governor. The rank of the adjutant general shall be at least that of brigadier general and the adjutant general shall hold office for a term of four years beginning and ending as provided in section 69.19. At the time of appointment the adjutant general shall be a federally recognized commissioned officer in the United States army or air force, the army or air national guard, the army or air national guard of the United States, or the United States army or air force reserve who has reached at least the grade of colonel and who is or is eligible to be federally recognized at the next higher rank.

Statutory Authority: Iowa Code 1 § 29C.5. A homeland security and emergency management division is created within the department of public defense.

Source: http://nxtsearch.legis.state.ia.us/NXT/gateway.dll/2007%20Iowa%20Code/2007code/1/2/2130/2131?f=templates$fn=defaultURLQueryLink.htm$q=[field%20folio-destination-name:'ch_29C']$x=Advanced (Chapter 29C—Emergency Management and Security).

Adjutant General

Director: Major General G. Ron Dardis

Functions/Responsibilities: Iowa Code 1 § 29.1. The department of public defense is composed of the military division and the homeland security and emergency management division. The adjutant general is the director of the department of public defense and the budget and personnel of all of the divisions are subject to the approval of the adjutant general. Iowa Code 1 § 29A.12. The adjutant general shall have command and control of the military division, and perform such duties as pertain to the office of the adjutant general under law and regulations, pursuant to the authority vested in the adjutant general by the governor. The adjutant general shall superintend the preparation of all letters and reports required by the United States from the state, and perform all the duties prescribed by law. The adjutant general shall have charge of the state military reservations, and all other property of the state kept or used for military purposes.

Qualifications: Iowa Code 1 § 29A.11. There shall be an adjutant general of the state who shall be appointed and commissioned by the governor subject to confirmation by the senate and who shall serve at the pleasure of the governor. The rank of the adjutant general shall be at least that of brigadier general and the adjutant general shall hold office for a term of four years beginning and ending as provided in section 69.19. At the time of appointment the adjutant general shall be a federally recognized commissioned officer in the United States army or air force, the army or air national guard, the army or air national guard of the United States, or the United States army or air force reserve who has reached at least the grade of colonel and who is or is eligible to be federally recognized at the next higher rank.

Statutory Authority: Iowa Code 1 § 29.2. There shall be within the department of public defense, as a division thereof, a state military agency which shall be styled and known as the "military division, department of public defense," with the adjutant general as the administrator thereof.

Source: http://nxtsearch.legis.state.ia.us/NXT/gateway.dll/2007%20Iowa%20Code/2007code/1/2/2130/2131?f=templates$fn=defaultURLQueryLink.htm$q=[field%20folio-destination-name:'ch_29C']$x=Advanced (Chapter 29A—Military Code).

KANSAS

Emergency Management & Homeland Security

Director: Major General Tod M. Bunting

Functions/Responsibilities: Kan. Stat. § 48–905a. The division of emergency preparedness within the office of the adjutant general is hereby abolished and there is hereby established within the office of the adjutant general a division of emergency management. To the extent provided in this act, all of the powers, duties and functions of such division of emergency preparedness are hereby transferred to and conferred and imposed upon the division of emergency management. The division of emergency management and the powers, duties and functions thereof shall be administered, by the adjutant general, who shall be the chief administrative officer thereof, under the supervision of the governor. Kan. Stat. § 48–907. For purposes of administering the division of emergency management functions, the adjutant general's powers and duties shall include: (a) To adopt, amend and repeal rules and regulations; (b) to cooperate with the advisory commission to the council of national defense through its division of state and local cooperation, or with any similar federal agencies hereafter created, and with any departments or other federal agencies engaged in defense or emergency management activities; (c) to cooperate with emergency management agencies or councils and similar organizations of other states (d) to supervise and direct investigations, and report to the governor with recommendations for legislation or other appropriate action as the adjutant general deems necessary, with respect to any type of activity or matter of public concern or welfare insofar as the same is or may be related to emergency management.

Qualifications/Statutory Authority: Kan. Stat. § 48–203. The governor shall appoint, subject to confirmation by the senate as provided in Kan. Stat. § 75–4315b, one adjutant general with the rank of major general, who shall be chief of staff. The person appointed shall have served at least five years as a commissioned officer in the Kansas national guard and shall have been an officer in the armed forces of the United States.

Source: http://www.kslegislature.org/legsrv-statutes/index.do (Chapter 48—Militia, Defense and Public Safety, Article 9—Emergency Preparedness for Disasters).

Adjutant General

Director: Major General Tod M. Bunting

Functions/Responsibilities: Kan. Stat. § 48–204. The adjutant general shall: (1) Be in control of the military department of the state and subordinate only to the governor in matters pertaining to the department; (2) have general supervision over all the subordinate military departments, to include the department of the army national guard and the department of the air national guard. (3) perform such duties as pertain to the adjutant general's department under the regulations and usage of the army of the United States; (4) superintend the preparation of all returns and reports required by the United States from the state.

Qualifications/Statutory Authority: Kan. Stat. § 48–203. The governor shall appoint, subject to confirmation by the senate as provided in K.S.A. 75–4315b, one adjutant general with the rank of major general, who shall be chief of staff. The person appointed shall have served at least five years as a commissioned officer in the Kansas national guard and shall have been an officer in the armed forces of the United States.

Source: http://www.kslegislature.org/legsrv-statutes/index.do (Chapter 48—Militia, Defense and Public Safety, Article 2—Kansas Army and Air National Guard).

KENTUCKY

Emergency Management

Director: Maxwell C. Bailey

Functions/Responsibilities: Ky. Rev. Stat. § 39A.070. The director, with the approval of the adjutant general, shall exercise the following powers, responsibilities, and duties: (1) To represent the Governor on all matters pertaining to the comprehensive emergency management program and the disaster and emergency response of the Commonwealth. (2) To coordinate the development of a statewide comprehensive emergency management program, and through it, an integrated emergency management system for the disaster and emergency response of the Commonwealth; (3) To promulgate administrative regulations and issue orders, directives, standards, rules, procedures, guidance, or recommended practices necessary to coordinate the development, administration, organization, operation, implementation, and maintenance of the statewide comprehensive emergency management program and the integrated emergency management system of the Commonwealth; (4) To coordinate the development of comprehensive emergency management programs by the cities, counties, and urban-county or charter county governments as functional components of the integrated emergency management system of the Commonwealth.

Qualifications: Ky. Rev. Stat. § 39A.210. No person shall be employed or associated in any capacity in any disaster and emergency response organization established under this chapter who advocates a change by force or violence in the constitutional form of the government of the United States or in this state or the overthrow of any government in the United States by force or violence, or who has been convicted of or is under indictment or information charging any subversive act against the United States. Each person who is appointed to serve in an organization for disaster and emergency response shall, before entering upon his or her duties, take an oath, in writing, before a person authorized to administer oaths in this Commonwealth.

Statutory Authority: Ky. Rev. Stat. § 39A.030. The Division of emergency Management is hereby created as the emergency management agency of the state government. § 39A.060 (1). The Division of Emergency Management shall be headed by a director recommended by the Adjutant General and appointed by the Governor.

Source: http://www.lrc.ky.gov/search.htm (Title V—Military Affairs, chapter 39A—Statewide Emergency Management Programs).

Department of Homeland Security

Director: Alecia Webb-Edgington

Functions/Responsibilities: Ky. Rev. Stat. § 39G.010. (a) Develop and publish a comprehensive statewide homeland security strategy that coordinates state and local efforts to detect, deter, mitigate, and respond to a terrorist incident; (b) Develop a comprehensive strategy addressing how state and federal funds and other assistance will be allocated within the state to purchase specialized equipment required to prevent and respond effectively and safely to terrorist incidents; (c) Urge the state and local governments to exceed minimum federal requirements for receiving assistance in preparing to respond to acts of war or terrorism in the hope that the Commonwealth will become a national leader in this preparation; (d) Provide information explaining how individuals and private organizations, including volunteer and religious organizations, can best prepare for and respond to incidents contemplated by this section and to other threatened, impending, or declared emergencies and whom to contact should they desire to volunteer help or services during such an emergency. The program shall identify and encourage these private organizations to specifically commit to provide food, shelter, personnel, equipment, materials, consultation, and advice, or other services needed to respond to these incidents.

Qualifications/Statutory Authority: Ky. Rev. Stat. § 39G.010. The Kentucky Office of Homeland Security shall be attached to the Office of the Governor and shall be headed by an executive director appointed by the Governor.

Source: http://law.justia.com/kentucky/codes/039g00/010.html (Title V, chapter 39G— Kentucky Office of Homeland Security).

Adjutant General

Director: Major General Donald C. Storm

Functions/Responsibilities: Ky. Rev. Stat. § 37.180. The Governor shall be commander in chief of the Kentucky active militia, and the adjutant general shall be the administrative officer and shall be responsible to the Governor for the proper functioning of the Kentucky active militia, and is hereby authorized and empowered to take necessary action to perfect and maintain an efficient organization for the purposes herein set out. KY Rev. Stat. § 36.040. (1) (a) Represent the Governor in all military matters pertaining to the Commonwealth of Kentucky; (b) Be the executive head of the Department of Military Affairs and exercise all functions vested by law in the department; (c) Establish the internal organizational structure of the major organizations of the department; (g) Direct and supervise the chiefs of staff departments and supervise all troops and all departments, arms, and branches of the Kentucky Army and Air National Guard.

Qualifications: Ky. Rev. Stat. § 36.020. The Governor, immediately on his induction into office, shall appoint the adjutant general who shall have served honorably, or be serving, as a commissioned officer of the Armed Forces of the United States, and who shall: (a) Have at least ten (10) years commissioned service in the Kentucky National Guard. A national guard active duty tour or mobilization of national guard units into active federal duty shall not be considered a break in national guard service. (b) Have attained at least the rank of lieutenant colonel with federal recognition. (c) Have not been separated from the Kentucky National Guard for more than five (5) years before the date of appointment. (d) Have met the federal recognition requirement for his current rank and be qualified to meet the requirements for federal recognition in the ranks of brigadier general and major general at the time of appointment to those ranks.

Statutory Authority: Ky. Rev. Stat. § 36.110. The staff of the commander in chief shall consist of an administrative and personal staff. The administrative staff shall consist of the adjutant general and such other officers of the Kentucky National Guard, of the grade prescribed by the Department of defense of the United States for the administrative staff for Kentucky, as are detailed by the governor.

Source: http://www.lrc.ky.gov/search.htm (Title V, chapter 36—Department of Military Affairs, and Chapter 37—Active Militia).

LOUISIANA

Office of Homeland Security and Emergency Preparedness

Director: Jeff Smith

Functions/Responsibilities: La. Rev. Stat. 29, § 725 H. The director, subject to the direction and control of the governor, shall be the executive head of the state homeland security and emergency preparedness agency and as such shall be responsible to the governor for carrying out the programs for homeland security and emergency preparedness for the state of Louisiana. He shall coordinate the activities of all agencies and organizations for homeland security and emergency preparedness within the state and shall maintain liaison with and cooperate with homeland security and emergency preparedness agencies and organizations of other states and of the federal government. La. Rev. Stat. 29, § 726 B. The office shall prepare and maintain a homeland security and state emergency operations plan and keep it current, which plan may include any of the following: (1) Prevention and minimization of injury and damage caused by disaster or emergency. (2) Prompt and effective response to disaster or emergency. (3) Emergency relief. (4) Identification of areas particularly vulnerable to disasters or emergency.

Qualifications: La. Rev. Stat. 29, § 725 C. (1) There shall be a director of the Governor's Office of Homeland Security and Emergency Preparedness who shall be appointed by the governor, subject to Senate confirmation. (2) The director shall have had at least ten years of emergency management experience or equivalent experience in emergency operations.

Statutory Authority: La. Rev. Stat. 29, § 725. The Governor's Office of Homeland Security and Emergency Preparedness is hereby established as a state agency within the office of the governor. There shall be a director of the Governor's Office of Homeland Security and Emergency Preparedness who shall be appointed by the governor, subject to Senate confirmation.

Source: http://law.justia.com/louisiana/codes/13/13.html (Title 29—Military, Naval and Veteran's Affairs, section 725—State Emergency and Disaster Agency).

Adjutant General

Director: Major General Bennett C. Landreneau

Functions/Responsibilities: La. Rev. Stat. 29, § 11. The adjutant general has control of the military department, subject to the orders of the governor, and performs the duties pertaining to the office of adjutant general under the laws of Louisiana and applicable federal law and regulations. The adjutant general shall have the authority to appoint, assign, promote, transfer, and separate all officers, including assistant adjutants general, in accordance with regulations promulgated and as provided in this Title. He shall superintend the preparation of all letters and reports pertaining to military affairs required by the United States of America from the state. He shall have charge of and is the responsible contracting authority for the state military reservations and all other state property kept, used, or operated by the military department.

Qualifications/Statutory Authority: La. Rev. Stat. 29, § 10. The adjutant general shall have a rank not lower than a brigadier general, and he shall have been a citizen of the state for at least fifteen years prior to his appointment. He shall be appointed by the governor with the consent of the Senate from active federally recognized officers of the Louisiana National Guard who have had at least seven years of federally recognized commissioned service in the Louisiana National Guard and have attained the federally recognized rank of colonel.

Source: http://www.legis.state.la.us/lss/lss.asp?doc=85330 (Title 29, chapters 10 and 11—Adjutant General).

MAINE

Emergency Management

Director: Robert McAleer

Functions/Responsibilities: The Maine Emergency Management Agency (MEMA) is responsible for homeland security and terrorism-related incidents, as well as natural diasters in Maine. MEMA is a bureau of the Department of Defense, Veterans, and Emergency Management. Me. Rev. Stat. 37–B, § 704. The director shall represent the Governor on all matters pertaining to the comprehensive emergency management program and the disaster and emergency response of the State. The director shall coordinate the activities of all organizations for emergency management within the State; shall maintain liaison with and cooperate with emergency management and public safety agencies and organizations of other states, the Federal Government and foreign countries, and their political subdivisions; prior to the annual meeting required in section 782, subsection 4, shall provide to each of the local emergency management organizations of the State an annual assessment of each organization's degree of emergency management capability and any other information pertinent to ensuring the public's welfare and safety within the local organization's jurisdiction; and has additional authority, duties and responsibilities as may be prescribed by the commissioner or the Governor. [2005, c. 634, § 11 (amd).]

Qualifications: Me. Rev. Stat. 37–B, § 704. The director must be qualified by education, training or experience in managing emergencies or in the emergency management profession and is appointed by the Governor upon recommendation of the commissioner, subject to review by the joint standing committee of the Legislature having jurisdiction over the Department of Public Safety and the Legislature. The director serves at the pleasure of the Governor. [2005, c. 634, § 11 (amd).]

Statutory Authority: Me. Rev. Stat. 37-B, § 704. The Maine Emergency Management Agency, as previously established and in this chapter called the "agency," is under the supervision of the Director of the Maine Emergency Management Agency, who in this chapter is called the "director."

Source: http://law.justia.com/maine/codes/title37-bch0sec0/title37-bch13sec0.html (Title 37–B—Defense, Veterans and Emergency Management, chapter 13—Maine Emergency Management Agency).

Department of Defense, Veterans, and Emergency Management

Director: Major General John Libby

Functions/Responsibilities: Me. Rev. Stat. 37–B, § 3. 1. The Adjutant General shall be the Commissioner of Defense, Veterans and Emergency Management. Me. Rev. Stat. 37–B, § 393. The Adjutant General operates the authority under the direction of the Governor and may: [2001, c. 374, § 8 (new).] 1. Execute agreements. Execute cooperative agreements between the Maine National Guard and the Federal Government or its instrumentalities or agencies; [2001, c. 374, § 8 (new).] 2. Contract with various entities. Contract with the Federal Government or its instrumentalities or agencies, the State or its agencies, instrumentalities or municipalities, foreign governments, public bodies, private corporations, partnerships, associations and individuals; [2001, c. 374, § 8 (new).] The Commissioner/Adjutant General serves as the Governor's official homeland security adviser.

Qualifications/Statutory Authority: Me. Rev. Stat. 37–B, § 107. A person appointed Adjutant General or assistant adjutant general must have attained the federally recognized rank of colonel in the Maine National Guard. [1999, c. 291, § 1 (rpr).]

Sources: http://law.justia.com/maine/codes/title37-bch0sec0/title37-bch0sec0.html (Title 37–B, chapter 1—Organization, and chapter 3—Military Bureau); and Web site for Homeland Security, State of Maine, http://www.maine.gov/mema/homeland/home_about.shtml.

MARYLAND

Emergency Management

Director: John Droneburg

Functions/Responsibilities: Md. Code § 14–104. The Director is the executive head of the Maryland Emergency Management Agency (MEMA). The Director is responsible to the Governor and the Adjutant General for carrying out the State emergency management program. If the Governor has formally declared the threat or occurrence of an emergency, the Director shall coordinate the activities of all organizations for emergency management operations in the State. With the approval of the Adjutant General and in collaboration with other public and private agencies in the State, the Director shall develop or cause to be developed mutual aid agreements for reciprocal emergency aid and assistance in case of emergency of an extreme nature that affects two or more political subdivisions.

Qualifications: Md. Code § 14–104. (a) The Adjutant General shall appoint the director of MEMA with the approval of the Governor. (b) The Director serves at the pleasure of the Adjutant General.

Statutory Authority: Md. Code. § 14–103. (a) There is a Maryland Emergency Management Agency in the Military Department. (b) The Director serves at the pleasure of the Adjutant General. (c) The Director is in the executive service of the State Personnel Management System and is entitled to the salary provided in the State budget.

Source: http://www.michie.com/maryland/lpext.dll?f=templates&fn=main-h.htm&cp=mdcode (Title 14, subtitle 1—Emergency Management Agency Act).

Governor's Office of Homeland Security

Director: Andrew Lauland

Functions/Responsibilities: Exec. Order 01.01.2003.18 (January 2003). The Director of the Governor's Office of Homeland Security will direct homeland security efforts across State Government and coordinate with federal and local government, private sector, academia, and the public to find solutions that ensure public safety while protecting individual freedoms; the Director will ensure that Maryland is a full and active partner in federal homeland security and will work to leverage federal initiatives and innovations to enhance public safety and security within the State and its subdivisions. The Director will: (1) Direct and coordinate homeland security activities within the state, paying special attention to key linkages between federal, State, regional, subdivision, metropolitan, and purely local activities; (2) Advise the Governor on policies, strategies, and measures to enhance and improve the ability to detect, prevent, prepare for, protect against, respond to, and recover from, man-made emergencies or disasters, including terrorist attacks; (3) Assess the homeland security of the State of Maryland on a continuing basis and ensure the development and updating of plans as appropriate.

Qualifications/Statutory Authority: Exec. Order 01.01.2003.18 (January 2003). The Governor's Office of Homeland Security is hereby established. The Office shall be managed by a Director. The Governor's Office of Homeland Security shall be comprised of limited professional staff appointed by the Governor. The Director shall serve at the will of the Governor.

Source: http://www.msa.md.gov/megafile/msa/speccol/sc5300/sc5339/000113/000000/000163/unrestricted/20040051e.html.

Adjutant General

Director: Major General Bruce F. Tuxill

Functions/Responsibilities: Md. Code § 13–301. (a) The Adjutant General: is the head of the Department. Md. Code § 13–302. (b) (1) The Adjutant General shall keep all records required to be kept and filed with the Adjutant General's office. (2) On or before each October 15, the Adjutant General shall submit to the Governor a detailed statement of all the receipts and expenditures for military purposes during the year ending the previous September 30. (c) (1) The Adjutant General is responsible for: (i) each armory that the State owns; and (ii) each building or other property purchased, occupied, or leased by or on behalf of the State military forces.

Qualifications/Statutory Authority: Md. Code § 13–301. (a) Shall be appointed by the Governor with the advice and consent of the Senate. (b) At the time of appointment, the Adjutant General shall: (1) have at least 10 years of commissioned field grade service in the National Guard; (2) have attained at least the rank of colonel; and (3) meet the requirements for federal recognition at the rank of Major General. (article IX—Militia and Military Affairs, section 2— Adjutant General).

Source: http://www.michie.com/maryland/lpext.dll?f=templates&fn=main-h.htm&cp=mode (Title 13—Militia, subtitle 3—adjutant general).

MASSACHUSETTS

Emergency Management

Director: Don Boyce

Functions/Responsibilities: MEMA Mission Statement: The Massachusetts Emergency Management Agency (MEMA) is the state agency responsible for coordinating federal, state, local, voluntary and private resources during emergencies and disasters in the Commonwealth of Massachusetts. MEMA provides leadership to: develop plans for effective response to all hazards, disasters or threats; train emergency personnel to protect the public; provide information to the citizenry; and assist individuals, families, businesses and communities to mitigate against, prepare for, and respond to and recover from emergencies, both natural and man made. (Source: http://mass.gov/?pageID=eopsterminal&L=3&L0=Home&L1=Public+Safety+Agencies&L2=M assachusetts+Emergency+Management+Agency&sid=Eeops&b=terminalcontent&f=mema_feat ure_story_2007_feature_stories_welcome_from_director_don_boyce&csid=Eeops (Executive Office of Public Safety and Security (EOPS) Web site).)

Qualifications: The governor appoints the head of MEMA. (Source: http://www.mass.gov/legis/ laws/mgl/31-3.htm.)

Statutory Authority: Gen. L. Chapter 6A § Section 18. The following state agencies are hereby declared to be within the executive office of public safety: the Massachusetts emergency management agency.

Source: http://www.mass.gov/legis/laws/mgl/6a-18.htm (Chapter 6A—Executive Offices, Section 181/2—Undersecretaries).

Homeland Security

Director: Juliette Kayyem

Functions/Responsibilities: Mass. Gen. L. 6A § 181/2. One undersecretary shall be the undersecretary of homeland security and shall oversee the functions and administration of the following boards and agencies: the emergency management agency, the department of fire services, the military department and the nuclear safety department. Each undersecretary shall conduct studies of the operations of each agency and work with each agency in effecting procedures and programs which promote efficiency and improvements in the administration of the agency. Each undersecretary shall assist the secretary in reviewing and acting upon budgetary and other financial matters concerning those agencies in accordance with sections 2C, 3, 3A, 4, 9B and 29 of chapter 29.

Qualifications/Statutory Authority: Mass. Gen. L. 6A § 181/2. The secretary shall, subject to section 3, appoint 4 undersecretaries. Each person appointed as an undersecretary shall have experience and shall know the field or functions of such position, shall receive such salary as the secretary shall determine and shall devote his full time to the duties of the office. One undersecretary shall be the undersecretary of homeland.

Source: http://www.mass.gov/legis/laws/mgl/6a-18.5.htm (Chapter 6A—Executive Offices, Section 181/2—Undersecretaries).

Adjutant General

Director: Brigadier General Oliver J. Mason Jr.

Functions/Responsibilities: Mass. Gen. L. 33 § 15. The adjutant general shall be charged with carrying out the policies of the commander-in-chief and shall issue orders in his name, but he shall not personally exercise command of troops in his capacity as adjutant general. He shall be the immediate adviser of the commander-in-chief on all matters relating to the militia and shall be charged with the planning, development and execution of the program of the military forces of the commonwealth. He shall cause the state staff to prepare plans for recruiting, organizing, supplying, equipping and mobilizing the organized militia, for use in the national defense, for state defense and emergencies, and for demobilizing the militia. He shall hold major organization commanders responsible for the proper training of their commands, and all orders and instructions for the government of the militia and of the officers and enlisted persons therein shall be issued and communicated to those concerned through military channels. Under the control of the governor, in his capacity as commander-in-chief, the adjutant general shall be the executive and administrative head of the military division of the executive branch of the government of the commonwealth. Except as otherwise provided, he shall require that all contracts and may require that all expenditures made by the division be submitted to him for approval.

Qualifications: Mass. Gen. L. 33 § 15. The adjutant general shall be appointed by the governor, in his capacity as commander-in-chief, from those persons who are, or have been, active commissioned officers in the Massachusetts national guard, army or air, for a period of not less than five years and who have attained, while serving therein, or in the national guard of the United States, a grade not lower than that of lieutenant colonel. He shall serve for a term coterminous with that of the governor and shall receive the same pay as an officer of the regular service of corresponding grade with corresponding length of service.

Statutory Authority: Mass. Gen. L. 33 § 15. The state staff shall consist of one adjutant general, with the grade of major general, who shall be the chief of staff to the commander-in-chief and the chief of the state staff, and the officers provided for in this section, each of whom shall perform his duties under the direction of the adjutant general.

Source: http://www.mass.gov/legis/laws/mgl/gl-33-toc.htm (chapter 33—Militia, section 15—State Staff).

MICHIGAN

Department of State Police

Director: Colonel Peter C. Munoz

Functions/Responsibilities: As Director of State Police, Colonel Munoz also serves as State Director of Emergency Management and as Michigan's Homeland Security Director. (Source: http://www.michigan.gov). Mich. Comp. Laws § 30.407. Sec. 7. The director shall implement the orders and directives of the governor in the event of a disaster or an emergency and shall coordinate all federal, state, county, and municipal disaster prevention, mitigation, relief, and recovery operations within this state. At the specific direction of the governor, the director shall assume complete command of all disaster relief, mitigation, and recovery forces, except the national guard or state defense force, if it appears that this action is absolutely necessary for an effective effort. The director's powers and duties shall include the administration of state and federal disaster relief funds and money; the mobilization and direction of state disaster relief forces; the assignment of general missions to the national guard or state defense force activated for active state duty to assist the disaster relief operations; the receipt, screening, and investigation of requests for assistance from county and municipal governmental entities; making recommendations to the governor; and other appropriate actions within the general authority of the director.

Qualifications: Appointed by the governor. (Source: http://www.michigan.gov/gov/0,1607,7-168-34750-110498--,00.html.)

Statutory Authority: Mich. Comp. Laws § 30.407a. The department shall establish an emergency management division for the purpose of coordinating within this state the emergency management activities of county, municipal, state, and federal governments. The division shall prepare and maintain a comprehensive Michigan emergency management plan that encompasses mitigation, preparedness, response, and recovery for the state.

Source: http://www.legislature.mi.gov/(S(jfov5n550e5ti055ecqjfgiy))/mileg.aspx?page=chapter index (chapter 30—Civilian Defense, section 407—Emergency Management Act).

Adjutant General

Director: Major General Thomas G. Cutler

Functions/Responsibilities: Mich. Comp. Laws § 32.700. Sec. 300. The office of the adjutant general, with the rank of major general in the national guard, is created. He shall be the commanding general of the military establishment. Under the direction of the governor he is charged with the responsibility for the command, administration, logistics, training and fiscal direction of the military establishment. He may perform any act authorized by this chapter or the regulations issued pursuant to this chapter through or with the aid of such officers, officials or directors of the military department as he may designate. The adjutant general shall direct the planning for the organization and employment of the forces of the organized militia in carrying out their state military mission and establish unified command of state forces whenever they shall be jointly engaged. Mich. Comp. Laws § 32.710. Sec. 310. The adjutant general is the military advisor to the governor and the director of the department of military and veterans affairs. The adjutant general's office is in Lansing. The adjutant general may publish orders and other directives in the name of the governor and this state to implement and administer the duties and responsibilities outlined in this act. The adjutant general's duties include the development and implementation of plans for the defense of state military personnel, lands, installations, and vital resources; maintenance of the personnel records of all active, inactive, retired, or deceased personnel of the state military establishment; and liaison in the transaction of official business for this state with the United States and with other states and territories, including those duties devolving upon the adjutant general pursuant to the national defense act and other pertinent federal laws and regulations.

Qualifications: Mich. Comp. Laws § 32.702. Sec. 302. The governor shall appoint the adjutant general from among qualified federally recognized officers of the national guard. The adjutant general shall have served as an officer of field or general grade in the state military establishment for not less than 5 years before appointment. The adjutant general shall serve at the pleasure of the governor, and unless sooner relieved, shall serve until the age of 64.

Statutory Authority: Mich. Comp. Laws § 32.700. Sec. 300. The office of the adjutant general, with the rank of major general in the national guard, is created.

Source: http://www.legislature.mi.gov/(S(elcr15q4cuhz13akuhlzi4bp))/mileg.aspx?page=Get MCLDocument&objectname=mcl-chap32 (chapter 32—Military Establishment, section 700—Michigan Military Act).

MINNESOTA

Department of Public Safety, Division of Homeland Security & Emergency Management

Director: Kris Eide

Functions/Responsibilities: Minn. Stat. § 12.09. *Coordination.* The Division of Emergency Management shall coordinate state agency preparedness for and emergency response to all types of natural and other emergencies and disasters, including discharges of oil and hazardous substances. *State emergency plan.* The division shall develop and maintain a comprehensive state emergency operations plan and emergency management program in accord with section 12.21, subdivision 3, clause (2), and ensure that other state emergency plans that may be developed are coordinated and consistent with the comprehensive state emergency operations plan. *State answering point system.* The division shall establish and maintain a single state answering point system for use by persons responsible for reporting emergency incidents and conditions involving hazardous substances or oil, nuclear power plant incidents or accidents, or other emergencies or disasters to state agencies, and for requesting state or federal assistance during and following an emergency or disaster. *Activation of emergency operations centers.* The division shall activate the state and regional emergency operations centers when an emergency or disaster threatens or has occurred. *Coordination of local programs.* The division shall coordinate the development and maintenance of emergency operations plans and emergency management programs by the political subdivisions of this state, with the plans and programs integrated into and coordinated with the emergency operations plan and emergency management program of this state to the fullest possible extent.

Qualifications/Statutory Authority: Minn. Stat. § 12.11. Subd. 1. A Division of Emergency Management is established within the Department of Public Safety under the supervision and control of the governor and a state director of emergency management. The commissioner of public safety shall appoint the state director, who shall not hold any other state office.

Source: http://ros.leg.mn/revisor/pages/search_text/doc_search.php?search=stat (chapter 12—Emergency Management Act).

Adjutant General

Director: Major General Larry W. Shellito

Functions/Responsibilities: Minn. Stat. § 190.09. Subd. 1. Official duties. The adjutant general shall be the chief of staff to the commander-in-chief and the administrative head of the Military Department. The adjutant general shall make an annual report to the governor, at such time as the governor may require, of all the transactions of the Military Affairs Department, setting forth the number, strength and condition of the National Guard, and such other matters as deemed important and shall make and transmit to the federal government the returns required by the laws of the United States. Subd. 2. Mission; efficiency. It is part of the department's mission that within the department's resources the adjutant general shall endeavor to: (1) prevent the waste or unnecessary spending of public money; (2) use innovative fiscal and human resource practices to manage the state's resources and operate the department as efficiently as possible; (3) coordinate the department's activities wherever appropriate with the activities of other governmental agencies.

Qualifications/Statutory Authority: Minn. Stat. § 190.07. There shall be an adjutant general of the state who shall be appointed by the governor. The adjutant general shall be a staff officer, who at the time of appointment shall be a commissioned officer of the National Guard of this state, with not less than ten years military service in the National Guard of this state or the armed forces of the United States, at least three of which shall have been commissioned and who shall have reached the grade of a field officer. The adjutant general shall hold at least the rank of major general and may be promoted to and including the highest rank authorized under federal law.

Source: http://ros.leg.mn/bin/getpub.php?pubtype=STAT_CHAP&year=2006&section=190 (chapter 190—Military Affairs).

MISSISSIPPI

Emergency Management

Director: Mike Womack

Functions/Responsibilities: Miss. Code § 33–15–7. The director, subject to the direction and control of the Governor, shall be the executive head of the emergency management agency and shall be responsible to the Governor for carrying out the program for emergency management of this state. He shall coordinate the activities of all organizations for emergency management within the state, and shall maintain liaison with and cooperate with emergency management agencies and organizations of other states and of the federal government, and shall have such additional authority, duties, and responsibilities authorized by this article as may be prescribed by the Governor.

Qualifications/Statutory Authority: Miss. Code § 33–15–7. There is hereby created within the executive branch of the state government a department called the Mississippi Emergency Management Agency with a director of emergency management who shall be appointed by the Governor; he shall hold office during the pleasure of the Governor and shall be compensated as determined by any appropriation that may be made by the Legislature for such purposes.

Source: http://michie.lexisnexis.com/mississippi/lpext.dll?f=templates&fn=main-h.htm&cp= (Title 33—Military Affairs, chapter 15—Emergency Management and Civil Defense).

Homeland Security

Director: J. W. Ledbetter

Functions/Responsibilities: Mississippi Homeland Security Web site: The backbone of Mississippi's Homeland Security lies in the Office of Homeland Security, which seeks to provide leadership in the deterrence, prevention, mitigation, and recovery of any and all terrorist activities and actions. (Source: http://www.mississippi.gov/frameset.jsp?URL=http%3A%2F%2Fwww.homelandsecurity.ms.gov%2F.)

Qualifications: Miss. Code § 45–1–2. The commissioner of the department shall appoint heads of offices, who shall serve at the pleasure of the commissioner. The commissioner shall have the authority to organize the offices established by subsection (2) of this section as deemed appropriate to carry out the responsibilities of the department. The organization charts of the department shall be presented annually with the budget request of the Governor for review by the Legislature.

Statutory Authority: Miss. Code § 45–1–2. The Commissioner of Public Safety shall establish the organizational structure of the Department of Public Safety, which shall include the creation of any units necessary to implement the duties assigned to the department and consistent with specific requirements of law, including, but not limited to: Office of Homeland Security.

Source: http://michie.lexisnexis.com/mississippi/lpext.dll?f=templates&fn=main-h.htm&cp= (Title 45—Public Safety and Good Order, chapter 1—Department of Public Safety).

Adjutant General

Director: Major General Harold A. Cross

Functions/Responsibilities: Miss. Code § 33–3–15. The Adjutant General, with the approval of the Governor, is expressly authorized to issue such orders, rules and regulations as may be necessary in order that the organization, training and discipline of the components of the militia of this state will at all times conform to the applicable requirements of the United States government relating thereto. Miss. Code § 33–3–11. [Through December 31, 2007, this section shall read as follows:] The Adjutant General shall: Appoint all of the employees of his department and he may remove any of them at his discretion; Submit to the Governor in each year preceding a regular session of the Legislature a printed detailed report of the transactions of his office, the expenses thereof, and such operations and conditions of the National Guard of this state as may be required by the Governor. The Adjutant General with the approval of the Governor shall provide for and be responsible for the organization, training, tactical employment, and discipline of the Mississippi National Guard, Mississippi State Guard, and the unorganized militia when called to active state duty. (Source: http://michie.lexisnexis.com/mississippi/lpext.dll?f=templates&fn=main-h.htm&cp=.)

Qualifications: Miss. Code § 33–3–7. To be eligible for such appointment, the Adjutant General shall have attained at least the rank of colonel, shall be eligible to receive federal recognition upon his appointment, and shall have served at least seven (7) years in the Armed Forces of the United States, either in active federal service or as a member of a reserve component, with at least three (3) years of such service in the Mississippi National Guard. At least five (5) years of such service shall have been as a commissioned officer. (Source: http://michie.lexisnexis.com/mississippi/lpext.dll?f=templates&fn=main-h.htm&cp=.)

Statutory Authority: Miss. Code § 33–3–3. Military department. There shall be in the executive branch of the state government a military department. The Adjutant General shall be the executive head of the department and, as such, subordinate only to the governor in matters pertaining thereto.

Source: http://michie.lexisnexis.com/mississippi/lpext.dll?f=templates&fn=main-h.htm&cp= (Title 33, chapter 3—Military Department and Adjutant General).

MISSOURI

Emergency Management

Director: Ronald M. Reynolds

Functions/Responsibilities: State Emergency Management Agency (SEMA) Mission Statement: "to protect the lives and property of all Missourians when major disasters threaten public safety in any city, county or region of Missouri. SEMA responds to two types of disasters—natural and manmade. Natural disasters are major snow and/or ice storms, floods, tornadoes and/or severe weather, as well as the threat of a serious earthquake along Missouri's New Madrid Fault. Manmade disasters, also known as technological emergencies, may include hazardous material incidents, nuclear power plant accidents and other radiological hazards. SEMA is also responsible for developing a State Emergency Operations Plan which coordinates the actions of Missouri State government departments and agencies in the event of any emergency requiring use of State resources and personnel." (Source: http://sema.dps.mo.gov/mission.htm.)

Qualifications: Mo. Rev. Stat. Title V, § 41.140. There shall be an adjutant general of the state appointed by the governor by and with the advice and consent of the senate who, at the time of his appointment, has not less than ten years of previous military service as a commissioned officer with the military forces of this state, or the United States, or in any or all of such services combined, five years of the service being in field grade. (Source: http://www.moga.mo.gov/ STATUTES/STATUTES.HTM (Title V—Military Affairs and Police, chapter 41—Military Forces).)

Statutory Authority: Mo. Rev. Stat. Title V, § 44.020. There is hereby created within the military division of the executive department, office of the adjutant general, the "State Emergency Management Agency", for the general purpose of assisting in coordination of national, state and local activities related to emergency functions by coordinating response, recovery, planning and mitigation. The adjutant general, subject to the direction and control of the governor, shall be the executive head of the office of the state emergency management agency.

Source: http://www.moga.mo.gov/STATUTES/STATUTES.HTM (Title V, chapter 44—Civil Defense).

Homeland Security Advisory Council

Director: Mark James

Functions/Responsibilities: Exec. Order 05–20 (July 2005). The Missouri Homeland Security Advisory Council is charged with the task of ensuring that proper homeland security plans and coordination are in place at the state and local level and that homeland security grant expenditures are done in a coordinated and efficient way. The council is tasked to: 1. Evaluate current state homeland security plans and make recommendations for modifications to ensure that Missouri's plans are properly equipped to respond to threats, whether foreseeable or unforeseeable. 2. Work with city and county officials to ensure that localities are integrated into and participating in homeland security planning and preparation. 3. Make recommendations for structural changes that will facilitate better cooperation and coordination between state and local homeland security/ emergency responder personnel.

Qualifications: Exec. Order 05–20 (July 2005). The Council shall consist of seventeen members. The Chairman of the Council shall be the Director of the Department of Public Safety.

Statutory Authority: Exec. Order 05–20 (July 2005). Matt Blunt, Governor of the State of Missouri, by virtue of the authority placed in him by the Constitution and the Laws of the State of Missouri, established the Missouri Homeland Security Advisory Council (herein called the "Council").

Source: http://www.gov.mo.gov/eo/2005/eo05_020.htm.

Adjutant General

Director: Major General King Sidwell

Functions/Responsibilities: Mo. Rev. Stat. § 41.160. The adjutant general shall be the military secretary and chief of staff to the commander in chief and the administrative head of the military establishment of the state. The adjutant general shall, under direction of the governor, be charged with the supervision of all matters pertaining to the administration, discipline, mobilization, organization and training of the organized militia of the state. The adjutant general shall supervise the preparation and submission of all returns and reports pertaining to the militia of the state as may be required by the United States. The adjutant general shall make an annual report to the governor at such time as the governor may require, setting forth the transaction of the office of the adjutant general, the strength and condition of the organized militia, including results of latest federal inspections, and such other matters as the adjutant general may deem important or may be requested by the governor.

Qualifications/Statutory Authority: Mo. Rev. Stat. § 41.140. There shall be an adjutant general of the state appointed by the governor by and with the advice and consent of the senate who, at the time of his appointment, has not less than ten years of previous military service as a commissioned officer with the military forces of this state, or the United States, or in any or all of such services combined, five years of the service being in field grade.

Source: http://law.justia.com/missouri/codes/t05/c041.html (Title V, chapter 41).

MONTANA

Department of Military Affairs, Division of Disaster & Emergency Services

Director: Dan McGowan

Functions/Responsibilities: Mont. Code. § 10–3–105. The department through the division of disaster and emergency services is responsible to the governor for carrying out the planning and program for disaster and emergency services of this state. The division shall prepare and maintain a comprehensive plan and program for disaster and emergency services of this state. The plan and program must be coordinated with the disaster and emergency plans and programs of the federal government, other states, political subdivisions, and Canada to the fullest extent possible.

Qualifications: Director's Position Description: Knowledge of concepts, theories, principles and practices of emergency management and related political process with a demonstrated ability to apply this knowledge. Bachelor's degree and extensive (6 or more years) professional experience or a master's degree and considerable experience.

Statutory Authority: Mont. Code. § 10–3–105. A division of disaster and emergency services is established in the department. The division must have an administrator and other professional, technical, secretarial, and clerical employees as necessary for the performance of its functions.

Source: http://law.justia.com/montana/codes/10/10_3_1.html (Title 10—Military Affairs and Disaster and Emergency Services, Chapter 3—Disaster and Emergency Services).

Homeland Security Task Force

Director: Dan McGowan

Functions/Responsibilities: Executive Order 28–1. I. With Disaster and Emergency Services (DES) as lead agency, the Task Force shall coordinate the following: the development of clear lines of communication and protocol for working with the U.S. Office of Homeland Security, the Anti-Terrorism Task Force as formed by the U.S. Department of Justice; and all other relevant federal, state, local, tribal and private agencies and organizations. With DES as lead agency, the Task Force shall: develop and implement a comprehensive statewide strategy to strengthen Montana's capabilities to detect, prepare for, prevent, protect against, respond to and recover from any terrorist threats or attacks within the state.

Qualifications: Executive Order 28–1. II. The Task Force shall be composed of the DES Administrator as Chairperson.

Statutory Authority: Executive Order 28–1. Judy Martz, Governor of the State of Montana, pursuant to the authority vested in her as Governor under the Constitution and laws of the State of Montana, established the Montana Homeland Security Task Force and designated DES as lead agency.

Source: http://dma.mt.gov/des/homelandsecurity/hsexecorder.asp.

Adjutant General

Director: Major General Randall D. Mosley

Functions/Responsibilities: Department of Military Affairs mission statement: Our Missions are: Federal—To serve as the primary federal reserve force in support of the national security objectives when called upon by the President of the United States. State—Protection of life property, preservation of peace, order and public safety for Montana's citizens, when called upon by the Governor. Community—Participate in local, state, and national programs that add value to America. (Source: http://dma.mt.gov/default.asp.) Mont. Code. § 2–15–1203. The adjutant general shall appoint, with the approval of the governor, an assistant adjutant general for the army national guard to be selected from the active list of the army national guard and an assistant adjutant general for the air national guard to be selected from the active list of the air national guard.

Qualifications: Mont. Code. § 2–15–1202. (1) (a) have the rank of major general; (b) be selected from the active list of the national guard of this state; (c) be federally recognized in the rank of lieutenant colonel or higher, immediately preceding his appointment; (d) have had at least 10 years of service as an officer of the active national guard of this state during the 15 years immediately preceding his appointment.

Statutory Authority: Mont. Code. § 2–15–1201. There is a department of military affairs. The department head is the adjutant general of the state.

Source: http://law.justia.com/montana/codes/2/2_15_12.html (Title 2—Government Administration, chapter 15, part 12—Department of Military Affairs).

NEBRASKA

Emergency Management

Director: Major General Roger P. Lempke

Functions/Responsibilities: Neb. Rev. Stat. § 81–829.41. The Nebraska Emergency Management Agency shall maintain an emergency operations plan and keep it current. Cooperate with the federal government and any public or private agency or entity in achieving any purpose of the act and in implementing programs for disaster prevention, mitigation, preparedness, response, and recovery and emergency management. Coordinate state emergency response as directed by the Governor. (Source: http://uniweb.legislature.ne.gov/ LegalDocs/view.php?page=s8108029041 (Title 81—State Administrative Departments).)

Qualifications: Neb. Rev. Stat. § 55–121. The Adjutant General shall be appointed by the Governor from the active or retired commissioned officers of the National Guard of this state. Such Adjutant General shall be or have been a commissioned officer who has actively served in the National Guard of this state for at least five years, shall have attained at least the grade of lieutenant colonel, and shall be able to become eligible for promotion to general officer. (Source: http://uniweb.legislature.ne.gov/LegalDocs/view.php?page=s5501021000 (Title 55—Militia).)

Statutory Authority: Neb. Rev. Stat. § 81–829.31. (1996) There is hereby created in the office of the Adjutant General the Nebraska Emergency Management Agency. The Adjutant General shall administer the Emergency Management Act.

Source: http://uniweb.legislature.ne.gov/ LegalDocs/view.php?page=s8108029031 (Title 81).

Homeland Security

Director: Carlos Castillo

Functions/Responsibilities: Neb. Rev. Stat. § 81–830. The purpose of the office is to ensure preparedness by the State of Nebraska in response to terrorist acts. The office shall coordinate efforts regarding domestic security issues with the United States Department of Homeland Security. The Director of State Homeland Security shall serve as the contact between the state and the United States Department of Homeland Security.

Qualifications/Statutory Authority: Neb. Rev. Stat. § 81–830. The Office of Homeland Security is created. The Governor shall appoint the Director of State Homeland Security who shall serve at the pleasure of the Governor.

Source: http://uniweb.legislature.ne.gov/LegalDocs/view.php?page=s8108030000 (Title 81).

Adjutant General

Director: Major General Roger P. Lempke

Functions/Responsibilities: Neb. Rev. Stat. § 55–122. The Adjutant General shall be in control of the military forces of the state and subordinate only to the Governor in matters pertaining to such forces. He shall issue and transmit all orders of the Governor with reference to the militia or military organization of the state, and shall keep a record of all officers commissioned by the

Governor and all general and special regulations, and of all such matters as pertain to the organization of the state militia and Nebraska National Guard. He shall have charge of, and receive and issue all ordnance and ordnance stores, clothing, camp, and garrison equipment, and other public property pertaining to the militia or National Guard of the state, and shall provide transportation and subsistence, when necessary, under authority of the Governor. (Source: http://uniweb.legislature.ne.gov/LegalDocs/view.php?page=s5501022000 (Title 55).)

Qualifications: Neb. Rev. Stat. § 55–121. The Adjutant General shall be appointed by the Governor from the active or retired commissioned officers of the National Guard of this state. Such Adjutant General shall be or have been a commissioned officer who has actively served in the National Guard of this state for at least five years, shall have attained at least the grade of lieutenant colonel, and shall be able to become eligible for promotion to general officer. (Source: http://uniweb.legislature.ne.gov/LegalDocs/view.php?page=s5501021000 (Title 55).)

Statutory Authority: Neb. Rev. Stat. § 81–830. Section 55–120 The Military Department shall consist of the Adjutant General in the grade of major general, one deputy adjutant general with a grade not less than colonel, one assistant adjutant general or chief of staff for Army National Guard affairs and one assistant adjutant general or chief of staff for Air National Guard affairs, each in the grade of brigadier general, one assistant director for Nebraska Emergency Management Agency affairs, and such other personnel as may be necessary to comply with such tables of organization as are prescribed for this state by the laws or regulations of the United States.

Source: http://uniweb.legislature.ne.gov/legaldocs/view.php?page=s55index (Title 81).

NEVADA

Department of Public Safety, Division of Emergency Management

Director: Frank Siracusa

Functions/Responsibilities: Nev. Rev. Stat. § 414.040. The Chief of the Division, subject to the direction and control of the Director, shall carry out the program for emergency management in this state. He shall coordinate the activities of all organizations for emergency management within the State, maintain liaison with and cooperate with agencies and organizations of other states and of the Federal Government for emergency management and carry out such additional duties as may be prescribed by the Director. The Chief shall assist in the development of comprehensive, coordinated plans for emergency management by adopting an integrated process, using the partnership of governmental entities, business and industry, volunteer organizations and other interested persons, for the mitigation of, preparation for, response to and recovery from emergencies or disasters.

Qualifications/Statutory Authority: Nev. Rev. Stat. § 414.040. A Division of Emergency Management is hereby created within the Department of Public Safety. The Chief of the Division is appointed by and holds office at the pleasure of the Director of the Department of Public Safety. The Division is the State Agency for Emergency Management and the State Agency for Civil Defense for the purposes of the Compact ratified by the Legislature pursuant to Nevada Revised Statutes 415.010.

Source: http://www.leg.state.nv.us/nrs/NRS-414.html#NRS414Sec038 (Title 36—Military Affairs and Civil Emergencies, chapter 414—Emergency Management).

Homeland Security

Director: Dale M. Carrison

Functions/Responsibilities: Nev. Rev. Stat. § 239C.160. The Nevada Commission on Homeland Security is tasked to: 1. Make recommendations to the Governor, the Legislature, agencies of this State, political subdivisions, businesses located within this State and private persons who reside in this State with respect to actions and measures that may be taken to protect residents of this State and visitors to this State from potential acts of terrorism and related emergencies. 2. Propose goals and programs that may be set and carried out, respectively, to counteract or prevent potential acts of terrorism and related emergencies before such acts of terrorism and related emergencies can harm or otherwise threaten residents of this State and visitors to this State. 3. (a) Identify and categorize such buildings, facilities, geographic features and infrastructure according to their susceptibility to and need for protection from acts of terrorism and related emergencies; and (b) Study and assess the security of such buildings, facilities, geographic features and infrastructure from acts of terrorism and related emergencies.

Qualifications/Responsibilities: Nev. Rev. Stat. § 239C.120. The Nevada Commission on Homeland Security is hereby created. The Governor or his designee shall: (a) Serve as Chairman of the Commission; and (b) Appoint a member of the Commission to serve as Vice Chairman of the Commission.

Source: http://www.leg.state.nv.us/NRS/NRS-239C.html#NRS239CSec160 (Title 19, chapter 239C—Homeland Security).

Adjutant General

Director: Major General Cynthia N. Kirkland

Functions/Responsibilities: Nev. Rev. Stat. § 412.048. The Adjutant General shall serve as the Chief of Staff to the Governor, the Director of the Office of the Military and the Commander of the Nevada National Guard, and: Is responsible, under the direction of the Governor, for the supervision of all matters pertaining to the administration, discipline, mobilization, organization and training of the Nevada National Guard, Nevada National Guard Reserve and volunteer organizations licensed by the Governor. Nev. Rev. Stat. § 412.052. The Adjutant General: Shall supervise the preparation and submission of all returns and reports pertaining to the militia of the State required by the United States. Is the channel of official military correspondence with the Governor, and, on or before November 1 of each even-numbered year, shall report to the Governor the transactions, expenditures and condition of the Nevada National Guard. The report must include the report of the United States Property and Fiscal Officer. Shall render such professional aid and assistance and perform such military duties, not otherwise assigned, as may be ordered by the Governor.

Qualifications: Nev. Rev. Stat. § 412.044. To be eligible for appointment to the office of Adjutant General, a person must be an officer of the Nevada National Guard, federally recognized in the grade of lieutenant colonel or higher, and must have completed at least 6 years' service in the Nevada National Guard as a federally recognized officer.

Statutory Authority: Nev. Rev. Stat. § 412.042. The military staff of the Governor consists of the Adjutant General, not more than two assistant adjutants general and personal aides-de-camp to the Governor selected from the commissioned officers of the Nevada National Guard or from reserve officers of the Armed Forces of the United States who are residents of Nevada and who are not serving on extended active duty.

Source: http://law.justia.com/nevada/codes/NRS-412.html#NRS412Sec042 (Title 36—Military Affairs and Civil Emergencies, chapter 412—State Militia).

NEW HAMPSHIRE

Homeland Security and Emergency Management

Director: Christopher M. Pope

Functions/Responsibilities: N.H. Rev. Stat. § 21–P:5–a. The director of homeland security and emergency management, under the supervision of the commissioner and the governor, shall devote full time and attention to overseeing the state-level planning, preparation, exercise, response to and mitigation of terrorist threats and incidents and natural and human-caused disasters. N.H. Rev. Stat. § 21–P:36. With general oversight by the assistant commissioner, the director shall coordinate the activities of all organizations for emergency management and emergency 911 telecommunications within the state, state and local, county, and private, and shall maintain liaison with and cooperate with police, fire, emergency medical and sheriff's departments and emergency management agencies and emergency telecommunications organizations of other states and of the federal government. (Source: http://law.justia.com/new hampshire/codes/nhtoc-i/21-p-36.html.)

Qualifications: N.H. Rev. Stat. § 21–P:5-a. Notwithstanding the provisions of RSA 21–G and RSA 21–P:3, the commissioner, after consultation with the governor, shall nominate for appointment by the governor and council, a director of homeland security and emergency management, who shall serve at the pleasure of the governor. The director of homeland security and emergency management shall be qualified by education and experience and shall receive the salary provided in RSA 94:1–a. (Source: http://law.justia.com/newhampshire/codes/nhtoc-i/21-p-5-a.html.)

Statutory Authority: N.H. Rev. Stat. § 21–P:36. There is hereby created a division of emergency services, communications, and management within the department of safety under the supervision of the director of emergency services, communications, and management and reporting to the assistant commissioner. The commissioner shall nominate a director of the division of emergency services, communications, and management, for appointment by the governor, with the consent of the council, and shall serve a term of 4 years until a successor has been appointed.

Source: http://www.gencourt.state.nh.us/rsa/html/i/21-p/21-p-36.htm (Title I—State Government, chapter 21–P—Department of Safety).

Adjutant General

Director: Major General Kenneth R. Clark

Functions/Responsibilities: N.H. Rev. Stat. § 110–B:8. The adjutant general shall be the chief of staff to the governor and shall be the executive head of the adjutant general's department. It shall be the duty of the adjutant general to direct the planning and employment of the forces of the national guard in carrying out their state military mission; to establish unified command of state forces whenever they shall be jointly engaged; to submit such written reports to the governor as the governor may prescribe; and to perform such other duties as the governor may direct. (Source: http://www.gencourt.state.nh.us/rsa/html/viii/110-b/110-b-mrg.htm.)

Qualifications: N.H. Rev. Stat. § 110–B:8. The adjutant general shall be appointed as provided in the constitution and the tenure of office shall be until the officer has reached the age of 65 years. At the time of appointment, the adjutant general shall have had not less than 5 years of service as a member of the New Hampshire national guard, immediately preceding that appointment, shall have attained at least the rank of colonel and shall be eligible for federal recognition by the department of defense as a brigadier general. (Source: http://law.justia.com/ newhampshire/codes/nhtoc-viii/110-b-7.html.)

Statutory Authority: N.H. Rev. Stat. § 110–B:7. The military staff of the governor shall consist of the adjutant general who shall be the chief of staff and 12 aides-de-camp.

Source: http://law.justia.com/newhampshire/codes/nhtoc-viii/110-b-7.html (Title VIII—Public Defense and Veterans Affairs, chapter 110—The Militia).

NEW JERSEY

State Police, Emergency Management Section

Section Supervisor: Major Richard Arroyo

Functions/Responsibilities: EMS Web site: The Emergency Management Section is the statewide agency with responsibility for planning, directing, and coordinating emergency operations within the state that are beyond local control. The section organizes, directs, staffs, coordinates and reports the activities of the Communications Bureau, Emergency Preparedness Bureau, and Recovery Bureau. The supervisor and staff facilitate the flow of information to and from the various bureaus supervised and serve as a conduit for communication with other division entities. (Source: http://www.njsp.org/divorg/homelandsec/ems.html.)

Qualifications: N.J. Stat. § 53:1–3. The superintendent may appoint a deputy superintendent with the rank of major. He shall receive such salary as shall be approved by the Attorney General and the president of the Civil Service Commission, subject to the availability of funds. (Source: http://law.justia.com/newjersey/codes/15d34/15d3a.html.)

Statutory Authority: Executive Order 39 (1954). The Office of Emergency Management is the lead State agency in disaster recovery operations and it is responsible for coordinating State preparedness plans for major disasters. Executive Order 101 (1980). Transfer of Emergency Management to the NJ State Police—Established an Office of Emergency Management in the Division of State Police, Department of Law and Public Safety. The Office of Emergency

Management shall be under the supervision, direction and control of the State Director of Emergency Management.

Source: http://www.state.nj.us/njoem/law_lawlist.html#enable.

State Police, Homeland Security Branch

Deputy Superintendent: Lieutenant Colonel Drew Lieb

Functions/Responsibilities: Mission statement: The mission of the branch is to provide a continuing preventive level of homeland security and public safety through the coordination of state resources. The goal is to provide for an increased capacity in responding to an elevation of the National Alert System and/or an event that necessitates additional mobilization of resources in concert with the state's law enforcement, intelligence, corporate, and emergency response partners. (Source: http://njsp.org/divorg/homelandsec/index.html)

Qualifications/Statutory Authority: N.J. Stat. § 53:1–3. The superintendent may appoint a deputy superintendent with the rank of major. He shall receive such salary as shall be approved by the Attorney General and the president of the Civil Service Commission, subject to the availability of funds. (Source: http://law.justia.com/newjersey/codes/15d34/15d3a.html.)

Office of Homeland Security and Preparedness

Director: Richard L. Cañas

Functions/Responsibilities: Executive Order #5 (March 16, 2006). Serve as the Homeland Security and Preparedness Advisor for the Governor and the state, and serve as the state's liaison with federal law enforcement authorities and with other states on counter-terrorism and emergency preparedness issues. Oversee, plan, and distribute discretionary state and federal funding for homeland security and emergency preparedness solely on the basis of risk, threat, and vulnerability. Ensure that the state has a comprehensive emergency plan. The Office shall be the central state agency responsible for the dissemination of counter-terrorism information/intelligence to federal, state, and local law enforcement entities, and for developing and administering training programs for law enforcement personnel. The Office shall also review and recommend modifications to all proposed state legislation regarding counter-terrorism and preparedness. (Source: http://www.state.nj.us/infobank/circular/eojsc5.htm) The state's Office of Counter-Terrorism and the Domestic Security Preparedness Task Force are part of the Office of Homeland Security and Preparedness. (Source: http://www.state.nj.us/njhomelandsecurity)

Qualifications/Statutory Authority: By executive order in 2006, Governor Jon Corzine created the Office of Homeland Security, which is empowered to administer, coordinate, lead, and supervise New Jersey's counter-terrorism and preparedness efforts. The goal of this Office is to coordinate emergency response efforts across all levels of government, law enforcement, emergency management, non-profit organizations, other jurisdictions, and the private sector, to protect the people of New Jersey. The Office shall be led by a Director, who shall report directly to the Governor and shall be a cabinet-level official. (Source: http://www.state.nj.us/infobank/circular/eojsc5.htm)

Adjutant General

Director: Major General Glenn K Rieth

Functions/Responsibilities: N.J. Stat. § 38A: 3–6. Exercise control over the affairs of the Department of Military and Veterans' Affairs and in connection therewith make and issue such regulations governing the work of the Department of Military and Veterans' Affairs and the conduct of its employees as may, in his judgment, be necessary or desirable. Command the organized militia of the State, with responsibility for recruiting, mobilization, administration, training, discipline, equipping, supply and general efficiency thereof. He may issue such regulations and delegate such command functions as he shall deem necessary. The regulations so issued shall, insofar as possible, conform to the federal laws and regulations concerning the same. Appoint and remove officers and other personnel employed within the department, subject to the provisions of N.J.S. 38A: 3–8 and Title 11A of the New Jersey Statutes and other applicable statutes, except as herein otherwise specifically provided. Have authority to organize and maintain an administrative division and to assign to employment therein secretarial, clerical and other assistants in the department or the Adjutant General's Office for the purpose of providing centralized support to all segments of the department, including budgeting, personnel administration and oversight of equal opportunity programs.

Qualifications: N.J. Stat. § 38A: 3–3. The head of the Department of Military and Veterans' Affairs shall be the Adjutant General, who shall be appointed with the grade of major general of the line, New Jersey Army National Guard, or major general, New Jersey Air National Guard, by the Governor, with the advice and consent of the Senate, from: (a) Federally recognized general officers in the national guard who have served therein for the preceding 10 years; or (b) Federally recognized commissioned officers in the national guard, who have served therein for the preceding 10 years and are now serving in a military grade not below that of a colonel, such officers having the qualifications to become federally-recognized as brigadier general of the line, New Jersey Army National Guard, or brigadier general, New Jersey Air National Guard.

Statutory Authority: N.J. Stat. § 38A: 2–2. The staff of the Governor shall consist of: (a) Executive: The Adjutant General of the State Department of Military and Veterans' Affairs.

Source: http://lis.njleg.state.nj.us/cgi-bin/om_isapi.dll?clientID=23644041&Depth=2&TD= RAP&advquery=%22adjutant%20general%22&depth=4&expandheadings=on&headingswithhit s=on&hitsperheading=on&infobase=statutes.nfo&rank=&record={DB78}&softpage=Doc_Fram e_PG42&wordsaroundhits=2&x=51&y=11&zz= (Title 38—Military and Veterans Law).

NEW MEXICO

Department of Homeland Security and Emergency Management

Director: Tim Manning

Functions/Responsibilities: N.M. Stat. Ann. § 9–28–2. The purpose of the Homeland Security and Emergency Management Department Act [9–28–1 NMSA 1978] is to establish a department to: A. consolidate and coordinate homeland security and emergency management functions to provide comprehensive and coordinated preparedness, mitigation, prevention, protection, response and recovery for emergencies and disasters, regardless of cause, and acts of terrorism. N.M. Stat. Ann. § 9–28–4. C. Except as otherwise provided in the Homeland Security and

Emergency Management Department Act [9–28–1 NMSA 1978], exercise general supervisory and appointing authority over all department employees, subject to any applicable personnel laws and rules; delegate authority to subordinates as the state director deems necessary and appropriate, clearly delineating such delegated authority and the limitations thereto; organize the department into those organizational units the state director deems will enable it to function most efficiently, subject to any provisions of law requiring or establishing specific organizational units; conduct research and studies that will improve the operations of the department and the provision of services to the residents of the state; provide courses of instruction and practical training for employees of the department and other persons involved in the administration of programs, with the objective of improving the operations and efficiency of administration.

Qualifications/Statutory Authority: N.M. Stat. Ann. § 9–28–4. A. The "homeland security and emergency management department" is created in the executive branch. The department is not a cabinet department. The chief administrative and executive officer of the department is the "state director of homeland security and emergency management", who shall be appointed by the governor and hold office at the pleasure of the governor.

Source: http://www.conwaygreene.com/nmsu/lpext.dll?f=templates&fn=main-hit-h.htm&2.0 (chapter 9—Executive Department, section 28—Homeland Security and Emergency Management Act).

Adjutant General

Director: Brigadier General Kenny C. Montoya

Functions/Responsibilities: N.M. Stat. Ann. § 20–3–2. B. The adjutant general is the military chief of staff to the governor and is the head of the department of military affairs. C. The adjutant general shall prescribe policies, rules and procedures for the orderly functioning of the department of military affairs which may include subordinate organizational structures and lines of authority. The adjutant general shall: A. prepare and publish, by order of the governor, such orders, rules and regulations, consistent with law, as are necessary to maintain the military forces in a state of efficiency in conformity with the needs of the state and the federal defense requirements; B. supervise the receipt, preservation, repair, distribution, issue and collection of all arms and military equipment of the state.

Qualifications: N.M. Stat. Ann. § 20–1–5. In case of a vacancy, the governor shall appoint as the adjutant general of New Mexico for a term of five years an officer who for three years immediately preceding his appointment as the adjutant general of New Mexico has been federally recognized as an officer in the national guard of New Mexico and who during his service in the national guard of New Mexico has received federal recognition in the rank of major or higher.

Statutory Authority: N.M. Stat. Ann. § 20–1–5. B. The adjutant general is the military chief of staff to the governor and is the head of the department of military affairs.

Source: http://www.conwaygreene.com/nmsu/lpext.dll?f=templates&fn=main-hit-h.htm&2.0 (chapter 9—Executive Department, section 9—Military Affairs).

NEW YORK

State Emergency Management Office

Director: John Gibb

Functions/Responsibilities: SEMO mission statement: The mission of the New York State Emergency Management Office (SEMO) is to protect the lives and property of the citizens of New York State from threats posed by natural or man-made events. (Source: http://www.semo. state.ny.us/about/.) N.Y. Exec. Law § 21–3. The disaster preparedness commission shall have the following powers and responsibilities: a. study all aspects of man-made or natural disaster prevention, response and recovery; b. request and obtain from any state or local officer or agency any information necessary to the commission for the exercise of its responsibilities; c. prepare state disaster preparedness plans, to be approved by the governor, and review such plans and report thereon by March thirty-first of each year to the governor, the legislature and the chief judge of the state. (Source: http://www.semo.state.ny.us/uploads/Article%202-B.pdf.)

Qualifications/Statutory Authority: N.Y. Exec. Law § 21–1. There is hereby created in the executive department a disaster preparedness commission. The governor shall designate the chair of the commission.

Source: http://law.justia.com/newyork/codes/executive/exc021_21.html.

Office of Homeland Security

Director: F. David Sheppard

Functions/Responsibilities: N.Y. Exec. Law § 26–709. 2. (a) oversee and coordinate the state's homeland security resources, subject to any laws, rules or regulations governing the budgeting and appropriation of funds. (b) review homeland security policies, protocols and strategies of state agencies; (c) develop policies, protocols and strategies, which may be used to prevent, detect, respond to and recover from terrorist acts or threats; (d) identify potential inadequacies in the state's policies, protocols and strategies to detect, respond to and recover from terrorist acts or threats; (e) undertake periodic drills and simulations designed to assess and prepare responses to terrorist acts or threats.

Qualifications: N.Y. Exec. Law § 26–710. The director of the office of homeland security director) shall be appointed by the governor, by and with the advice and consent of the senate, and hold office at the pleasure of the governor.

Statutory Authority: N.Y. Exec. Law § 26–729. There is hereby created within the executive department the office of homeland security, which shall have and exercise the powers and duties set forth in this article. Any reference to the 'office of public security' in the laws of New York state, executive orders, or contracts entered into on behalf of the state shall be deemed to refer to the state office of homeland security.

Source: http://public.leginfo.state.ny.us/menugetf.cgi?COMMONQUERY=LAWS (Article 26—State Office of Homeland Security).

Adjutant General

Director: Major General Joseph J. Taluto

Functions/Responsibilities: N.Y. Mil. Law § 1–11. The adjutant general shall exercise control over the division of military and naval affairs of the executive department of the state. It shall be the duty of the adjutant general to direct the planning and employment of the forces of the organized militia in carrying out their state military mission; to establish unified command of state forces whenever they shall be jointly engaged; to act as the state director of civil defense for the state; to submit an annual written report to the governor in such form as the governor may prescribe; and to perform such other duties as the governor may direct. The adjutant general may appoint such assistant adjutants general as may be necessary. Whenever the governor and those who would act in succession to the governor under the constitution and laws of the state shall be unable to perform the duties of commander-in-chief, the adjutant general shall command the militia.

Qualifications: N.Y. Mil. Law § 1–11. The adjutant general shall serve as such at the pleasure of the governor.

Statutory Authority: N.Y. Mil. Law § 1–10. The military staff of the governor shall consist of the adjutant general and such aides as the governor shall deem necessary who shall be commissioned officers detailed by the governor from the organized militia.

Source: http://public.leginfo.state.ny.us/menugetf.cgi (Military Law, article 1—The Militia of the State).

NORTH CAROLINA

Department of Crime Control and Public Safety, Division of Emergency Management & Homeland Security

Director: H. Douglas Hoell, Jr.

Functions/Responsibilities: N.C. Gen. Stat. § 166A–5. The Secretary of Crime Control and Public Safety shall be responsible to the Governor for State emergency management activities. To activate the State and local plans applicable to the areas in question and to authorize and direct the deployment and use of any personnel and forces to which the plan or plans apply, and the use or distribution of any supplies, equipment, materials and facilities available pursuant to this Article or any other provision of law. To develop a system of damage assessment through which the Secretary will recommend the appropriate level of disaster declaration to the Governor. Making of such studies and surveys of the resources in this State as may be necessary to ascertain the capabilities of the State for emergency management, maintaining data on these resources, and planning for the most efficient use thereof. (Source: http://www.ncga. state.nc.us/EnactedLegislation/Statutes/HTML/ByArticle/Chapter_166A/Article_1.html (chapter 166A—North Carolina Emergency Management Act).)

Qualifications: N.C. Gen. Stat. § 143B–9. The head of each principal State department, except those departments headed by popularly elected officers, shall be appointed by the Governor and serve at his pleasure. (Source: http://www.ncga.state.nc.us/enactedlegislation/statutes/html/by chapter/chapter_143b.html (chapter 143B—Executive Organization Act of 1973).)

Statutory Authority: N.C. Gen. Stat. § 166A–5. Secretary of Crime Control and Public Safety. – The Secretary of Crime Control and Public Safety shall be responsible to the Governor for State emergency management activities.

Source: http://www.ncga.state.nc.us/EnactedLegislation/Statutes/HTML/ByArticle/Chapter_166A/Article_1.html.

Adjutant General

Director: Major General William E. Ingram Jr.

Functions/Responsibilities: N.C. Gen. Stat. § 127A–20. In all administrative and operational matters affecting the militia while under State control, the Adjutant General shall be responsible to and subject to the direction and supervision of the Secretary of Crime Control and Public Safety. N.C. Gen. Stat. § 127A–19. The Adjutant General may appoint a deputy adjutant general for Army National Guard, an assistant adjutant general for Army National Guard, and an assistant adjutant general for Air National Guard, each of whom may hold the rank of brigadier general and who shall serve at the pleasure of the Governor.

Qualifications/Statutory Authority: N.C. Gen. Stat. § 127A–19. The military head of the militia shall be the Adjutant General who shall hold the rank of major general. The Adjutant General shall be appointed by the Governor in his capacity as commander in chief of the militia, in consultation with the Secretary of Crime Control and Public Safety, and shall serve at the pleasure of the Governor. No person shall be appointed as Adjutant General who has less than five years' commissioned service in an active status in any component of the armed forces of the United States. The Adjutant General, while holding such office, may be a member of the active national guard or naval militia.

Source: http://www.ncga.state.nc.us/enactedlegislation/statutes/html/bychapter/chapter_127a.html (chapter 127A—Militia).

NORTH DAKOTA

Department of Emergency Services, Division of Homeland Security

Director: Major General David Sprynczynatyk

Functions/Responsibilities: N.D. Cent. Code § 37–17.1–02.1. The department of emergency services consists of a division of state radio communications and a division of homeland security. The adjutant general serves as director, and is chairman of the policy-making Department of Emergency Services Advisory Committee. Reduce vulnerability of people and communities of this state to damage, injury, and loss of life and property resulting from natural or manmade disasters or emergencies, threats to homeland security, or hostile military or paramilitary action. (Source: http://www.legis.nd.gov/cencode/t37c171.pdf (Title 37—Military, chapter 17.1—Emergency Services).)

Qualifications: N.D. Cent. Code § 37–03–01. The governor shall appoint the adjutant general. Each candidate for the office must have been a federally recognized commissioned officer of the national guard for a period of at least three years immediately preceding the appointment, must have obtained the rank of lieutenant colonel or higher, and must have completed the educational requirements for appointment as a federally recognized general officer. (Source: http://www. legis.nd.gov/cencode/t37c03.pdf.)

Statutory Authority: N.D. Cent. Code § 37–17.1–02.1. The adjutant general is the director of this department.

Source: http://www.legis.nd.gov/cencode/t37c171.pdf.

Adjutant General

Director: Major General David A. Sprynczynatyk

Functions/Responsibilities: N.D. Cent. Code § 37–03–05. The adjutant general is in active control of the military department of this state and shall: 1. Perform the duties pertaining to the adjutant general and other chiefs of staff departments under the regulations and customs for the United States Army. 2. Superintend the preparation of all military returns and reports required by the United States from this state. 3. Keep a register of all the officers of the militia and national guard of this state. 4. Keep in the office of the adjutant general all records and papers required to be kept and filed in the office.

Qualifications: N.D. Cent. Code § 37–03–01. The governor shall appoint the adjutant general. Each candidate for the office must have been a federally recognized commissioned officer of the national guard for a period of at least three years immediately preceding the appointment, must have obtained the rank of lieutenant colonel or higher, and must have completed the educational requirements for appointment as a federally recognized general officer.

Statutory Authority: N.D. Cent. Code § 37–03–05. The Adjutant General is in active control of the military department of this state.

Source: http://www.legis.nd.gov/cencode/t37c03.pdf (Title 37, chapter 3—Adjutant General).

OHIO

Emergency Management

Director: Nancy Dragani

Functions/Responsibilities: OEMA Web site: The Ohio Emergency Management Agency is the central point of coordination within the state for response and recovery to disasters. Source: http://www.ema.ohio.gov/oema.asp. Ohio Rev. Code § 5502.22. The director, with the concurrence of the governor, shall appoint an executive director, who shall be head of the emergency management agency. The executive director may appoint a chief executive assistant, executive assistants, and administrative and technical personnel within that agency as may be necessary to plan, organize, and maintain emergency management adequate to the needs of the state. The executive director shall coordinate all activities of all agencies for emergency management within the state, shall maintain liaison with similar agencies of other states and of the federal government, shall cooperate with those agencies subject to the approval of the governor, and shall develop a statewide emergency operations plan that shall meet any applicable federal requirements for such plans.

Qualifications/Statutory Authority: Ohio Rev. Code § 5502.22. There is hereby established within the department of public safety an emergency management agency, which shall be governed under rules adopted by the director of public safety under section 5502.25 of the Revised Code. The director, with the concurrence of the governor, shall appoint an executive director, who shall be head of the emergency management agency.

Source: http://codes.ohio.gov/orc/5502.22 (Title 55—Department of Public Safety, section 5502.22—Emergency Management Agency).

Homeland Security

Director: William F. Vedra, Jr.

Functions/Responsibilities: Ohio Rev. Code § 5502.03. (B) (1) Coordinate all homeland security activities of all state agencies and be the liaison between state agencies and local entities for the purposes of communicating homeland security funding and policy initiatives. (2) Collect, analyze, maintain, and disseminate information to support local, state, and federal law enforcement agencies, other government agencies, and private organizations in detecting, deterring, preventing, preparing for, responding to, and recovering from threatened or actual terrorist events. This information is not a public record pursuant to section 149.43 of the Revised Code. (3) Coordinate efforts of state and local governments and private organizations to enhance the security and protection of critical infrastructure and key assets in this state; (4) Develop and coordinate policies, protocols, and strategies that may be used to prevent, detect, prepare for, respond to, and recover from terrorist acts or threats; (5) Develop, update, and coordinate the implementation of an Ohio homeland security strategic plan that will guide state and local governments in the achievement of homeland security in this state.

Qualifications: Ohio Rev. Code § 5502.03. The director of public safety shall appoint an executive director, who shall be head of the division of homeland security and who regularly shall advise the governor and the director on matters pertaining to homeland security. The executive director shall serve at the pleasure of the director of public safety.

Statutory Authority: Ohio Rev. Code § 5502.03. There is hereby created in the department of public safety a division of homeland security.

Source: http://codes.ohio.gov/orc/5502 (Title 55, chapter 5502.03—Division of Homeland Security).

Adjutant General

Director: Major General Gregory L. Wayt

Functions/Responsibilities: Ohio Rev. Code § 5913.07. The Adjutant General is the administrative head of the Ohio organized militia. Ohio Rev. Code § 5913.01. Keep and preserve the arms, ordnance, equipment, and all other military property belonging to the state or issued to the state by the federal government and issue any regulations necessary to keep, preserve, and repair the property as conditions demand. Submit an annual report to the governor at such time as the governor requires of the transaction of the adjutant general's department, setting forth the strength and condition of the Ohio organized militia and other matters that the adjutant general chooses. Command the state area command of the Ohio national guard.

Qualifications: Ohio Rev. Code § 5913.021. The adjutant general at the time of appointment shall be a federally recognized officer in the Ohio national guard in the grade of colonel or above. The adjutant general, the assistant adjutant general for army, the assistant adjutant general for air, and the assistant quartermaster general at the time of appointment shall each have not less than ten years' commissioned service in the armed forces of the United States, not less than five years of that service being in the Ohio national guard, and shall at all times during their tenure of office be federally recognized officers of the Ohio national guard.

Statutory Authority: Ohio Rev. Code § 5913.01. The adjutant general is the administrative head of the Ohio organized militia.

Source: http://codes.ohio.gov/orc/5913 (Title 59—Veterans-Military Affairs, chapter 5913—Adjutant General).

OKLAHOMA

Emergency Management

Director: Albert Ashwood

Functions/Responsibilities: Okla. Stat. § 63–683.4.D. The Director, subject to the direction and control of the Governor, shall be the executive head of the Department. The Director shall: coordinate the activities of all organizations for civil defense within the state; maintain liaison with and cooperate with the civil defense agencies and organizations of other states and of the federal government; develop and maintain a comprehensive mitigation plan for this state. The Director shall supervise the formulation, execution, review and revisions of the Emergency Resources Management Plan.

Qualifications/Statutory Authority: Okla. Stat. § 63–683.4.A. There is hereby created the Oklahoma Department of Civil Emergency Management. The Governor shall appoint a Director of the Department, with the advice and consent of the Senate, who shall be the head of the Department.

Source: http://oklegal.onenet.net/oklegal-cgi/get_statute?99/Title.63/63-683.4.html (Title 63—Public Health and Safety, section 683—Oklahoma Emergency Management Act).

Homeland Security

Director: Kerry Pettingill

Functions/Responsibilities: Okla. Stat. § 74–51.1. The Oklahoma Office of Homeland Security shall have the following duties: Establish a plan for the effective implementation of a statewide emergency All-Hazards response system, including the duties and responsibilities of regional emergency response teams. The Oklahoma Homeland Security Director shall have the duty and responsibility to develop and coordinate the implementation and administration of a comprehensive statewide strategy to secure the State of Oklahoma from the results of acts of terrorism, from a public health emergency, from cyberterrorism, and from weapons of mass destruction as that term is defined in 18 U.S.C., Section 2332a, and to perform other duties assigned by the Governor. Coordinating the Homeland Security efforts within the State of Oklahoma, including working with the Governor and Legislature, state agencies, and local elected officials and local governments, emergency responder groups, private-sector businesses, educational institutions, volunteer organizations, and the general public.

Qualifications/Statutory Authority: Okla. Stat. § 74–51.1. There is hereby created the Oklahoma Office of Homeland Security. The Governor shall be the chief officer of the Office and shall appoint a Homeland Security Director who shall be responsible to the Governor for the operation and administration of the Office. The Governor shall determine the salary for the Director. The Oklahoma Homeland Security Director shall possess or obtain a federally recognized Top Secret Level Clearance.

Source: http://www.lsb.state.ok.us/ (Title 74—State Government, section 51—Oklahoma Homeland Security Act).

Adjutant General

Director: Major General Harry M. Wyatt III

Functions/Responsibilities: Okla. Stat. § 44–21. The Military Department of the State of Oklahoma is hereby established and shall be administered and controlled by the Governor as Commander in Chief, with the Adjutant General as the executive and administrative head thereof. (Source: http://oklegal.onenet.net/oklegal-cgi/get_statute?99/Title.44/44-21.html.) Okla. Stat. § 44–26. The Adjutant General shall be in control of the Military Department of the state, subordinate only to the Governor, whose military adviser he shall be. Within the limitations and under the provisions of law, he shall supervise and direct the National Guard within the service of the state and when under state control in all of its organization, training and other activities; shall receive and give effect to the orders of the Governor; and shall perform such other military and defense duties, not otherwise assigned by law, as the Governor may prescribe. (Source: http://oklegal.onenet.net/oklegal-cgi/get_statute?99/Title.44/44-26.html.)

Qualifications/Statutory Authority: Okla. Stat. § 44–24. The Adjutant General shall be appointed by the Governor, by and with the advice and consent of the Senate, and shall serve during the pleasure of the Governor. No person shall be eligible to hold the office of the Adjutant General of this state, unless, at the time of his appointment, he is a federally recognized officer of the National Guard of Oklahoma, and of the National Guard of the United States, not below the rank of Major, and that his status as a federally recognized officer, both of the National Guard of Oklahoma and of the National Guard of the United States, shall have existed for at least three (3) years prior to the time of such appointment; or unless, within five (5) years prior to the time of his appointment, he has been a federally recognized officer of the National Guard of Oklahoma, and of the National Guard of the United States, not below the rank of Major, and that during his military service he served for a period of three (3) years as a federally recognized officer, both of the National Guard of Oklahoma and of the National Guard of the United States; provided that if the National Guard of Oklahoma is in active federal service and no persons having the above qualifications are available within the state, then the Governor may appoint, subject to the advice and consent of the Senate, any suitably qualified person who at any time in the preceding ten (10) years would have been qualified, as above, and who has served at least two (2) years in active federal service in the grade of Major or higher. (Source: http://oklegal.onenet.net/oklegal-cgi/get_statute?99/Title.44/44-24.html.)

Source: http://oklegal.onenet.net/oklegal-cgi/isearch (Title 44—Militia, section 24—Oklahoma Military Code).

OREGON

Emergency Management

Director: Ken Murphy

Functions/Responsibilities: Or. Rev. Stat. § 401.270. The Director of the Office of Emergency Management shall be responsible for coordinating and facilitating emergency planning, preparedness, response and recovery activities with the state and local emergency services agencies and organizations. Make rules that are necessary and proper for the administration and implementation of ORS 401.015 to 401.105, 401.260 to 401.325, 401.355 to 401.580 and 401.706; Coordinate the activities of all public and private organizations specifically related to providing emergency services within this state; Maintain a cooperative liaison with emergency management agencies and organizations of local governments, other states and the federal government; Serve as the Governor's authorized representative for coordination of certain response activities and managing the recovery process; Establish training and professional standards for local emergency program management personnel.

Qualifications/Statutory Authority: Or. Rev. Stat. § 401.260. The Emergency Management Division that has operated under this chapter is continued as the Office of Emergency Management within the Department of State Police and is made the emergency management agency for the State of Oregon. The office shall be under the supervision of a director appointed by the Superintendent of State Police with the approval of the Governor. The appointee shall serve at the pleasure of the superintendent, shall not be subject to the State Personnel Relations Law and shall be qualified by education, training and experience in the emergency management profession.

Source: http://www.leg.state.or.us/ors/401.html (chapter 401—Emergency Services and Communications, sections 260 and 270, Office of Emergency Management).

Homeland Security

Director: Ken Murphy

Functions/Responsibilities: Office of Homeland Security Mission Statement: To provide leadership in the protection of Oregonians, their property, the environment, and Oregon's economy from disasters and acts of terrorism through prevention, preparedness, responsiveness and recovery efforts. (Source: http://www.oregon.gov/OOHS/about_us.shtml.) Exec. Order 04–05 (2004). The Office of Homeland Security shall provide overall leadership in coordinating private and governmental sector efforts to prevent, prepare for, respond to, and recover from, disasters and terrorist attacks making efficient and effective use of Oregon's emergency resources, including close cooperation with the State's local partners and first responders.

Qualifications/Statutory Authority: Exec. Order 04–05 (2004). The Office of Homeland Security is hereby created. The Office of Homeland Security shall be administratively deemed a department, and shall perform the statutory duties required of each of its composite elements. The Director of Homeland Security shall be appointed by the Governor and shall be a member of the Governor's Cabinet.

Source: http://www.oregon.gov/Gov/pdf/ExecutiveOrder04-05.pdf.

Adjutant General

Director: Major General Raymond F. Rees

Functions/Responsibilities: Or. Rev. Stat. § 396.145. The military staff of the Governor shall consist of the Chief of Staff to the Governor, the Military Council and such personal aides-de-camp as the Governor shall deem necessary. The Adjutant General shall be Chief of Staff to the Governor. Or. Rev. Stat. § 396.160. The Adjutant General shall be the Director of the Oregon Military Department, and Chief of Staff to the Governor. The Adjutant General shall be the Commander of the Oregon National Guard. The Adjutant General shall be charged, under the direction of the Governor, with the supervision of all matters pertaining to the administration, discipline, mobilization, organization and training of the Oregon National Guard and the Oregon State Defense Force. The Adjutant General shall be the channel of official military correspondence with the Governor, and shall, on or before November 1 of each year, make a report to the Governor of the transactions, expenditures and condition of the Oregon National Guard. The report shall include the report of the United States Property and Fiscal Officer.

Qualifications/Statutory Authority: Or. Rev. Stat. § 396.150. (1) The Governor shall appoint an Adjutant General who shall hold office for a four-year term or until relieved by reason of resignation, withdrawal of federal recognition or for cause to be determined by a court-martial. (2) To be eligible for appointment to the office of Adjutant General, a person must be an officer of the Oregon National Guard, federally recognized in the grade of lieutenant colonel or higher, and must have completed at least six years' service in the Oregon National Guard as a federally recognized officer.

Source: http://law.justia.com/oregon/codes/vol10/396.html (Title 32—Military Affairs; Emergency Services, chapter 396—Militia Generally).

PENNSYLVANIA

Emergency Management Agency

Director: Robert French

Functions/Responsibilities: Pa. Code 35 § 7313. To prepare, maintain and keep current a Pennsylvania Emergency Management Plan for the prevention and minimization of injury and damage caused by disaster, prompt and effective response to disaster and disaster emergency relief and recovery. (Source: http://www.legis.state.pa.us/WU01/LI/LI/CT/35.HTM.)

Qualifications: Pa. Code 35 § 7313. To establish, equip and staff a Commonwealth and area emergency operations center with a consolidated Statewide system of warning and provide a system of disaster communications integrated with those of Federal, Commonwealth and local agencies involved in disaster emergency operations. To promulgate, adopt and enforce such rules, regulations and orders as may be deemed necessary to carry out the provisions of this part. To provide emergency direction and control of Commonwealth and local emergency operations. (Source: http://www.legis.state.pa.us/WU01/LI/LI/CT/35.HTM.)

Statutory Authority: Pa. Code 35 § 7312 (e) State director. – To supervise the work and activities comprising the State Civil Defense and Disaster Program, the Governor shall appoint an individual to act, on a full-time basis, as director of the agency.

Source: http://www.legis.state.pa.us/WU01/LI/LI/CT/35/00.073..HTM (Title 35—Health and Safety, chapter 73—Commonwealth Services, subchapter B—Pennsylvania Emergency Management Agency).

Homeland Security

Director: James F. Powers, Jr.

Functions/Responsibilities: Pennsylvania Homeland Security Web site: The Commonwealth's primary point-of-contact on homeland security issues and the Governor's senior advisor on homeland security issues. (Source: http://www.homelandsecurity.state.pa.us/homelandsecurity/ cwp/view.asp?a=445&Q=175059&homelandsecurityNav=|7312|.) Mission: Manage the Commonwealth's overall Protection framework and oversee implementation and continual evaluation of the Commonwealth Critical Infrastructure Protection Program. This comprises a 5-fold mission: Identify critical infrastructure (CI), key resources (KR), and significant special events (SSI). Assess Risks (Consequences, Vulnerabilities, & Threats). Determine the gaps in capabilities to respond. Partner with Federal and Commonwealth agencies and municipality, county, and private sector entities to develop strategies to mitigate the risk. Prioritize and implement risk-based protective programs. (Source: http://www.homelandsecurity.state.pa.us/ homelandsecurity/cwp/view.asp?a=378&q=175241.)

Qualifications: Appointed by the governor. (Source: http://www.homelandsecurity.state.pa.us/ homelandsecurity/ cwp/view.asp?a=445&Q=175059&homelandsecurityNav=|7312|.)

Source: http://www.homelandsecurity.state.pa.us/homelandsecurity/cwp/view.asp?a=445&Q= 175059&homelandsecurityNav=|7312|.

Adjutant General

Director: Major General Jessica L. Wright

Functions/Responsibilities: Pa. Code 9 § 902. The Adjutant General as head of the department is responsible to the Commonwealth and to the Governor for the organization and functioning of said department, and the performance and carrying out of all the duties, powers and responsibilities given or delegated. Maintain armories, arsenals, military reservations, air bases and all property and equipment intended for the use and training of the Pennsylvania military forces. Maintain a list of active and retired members of the Pennsylvania National Guard with name, rank, organization, date of appointment, date of retirement and residence. Execute and enforce the policies of the Commonwealth relative to the Pennsylvania military forces. (Source: http://www.legis.state.pa.us/WU01/LI/LI/CTS/51/00.009.002.000..HTM.)

Qualifications: Pa. Code 9 § 901. No Adjutant General, Deputy Adjutant General or Assistant Adjutant General shall be appointed who shall not have served at least ten years as a commissioned officer in the Pennsylvania National Guard, or any of the armed forces of the United States or their reserve components; the afore said service may be cumulative.

Statutory Authority: Pa. Code 9 § 901. The Governor shall appoint the Adjutant General with the advice and consent of the Senate.

Source: http://www.legis.state.pa.us/WU01/LI/LI/CTS/51/00.009.001.000..HTM (Title 51—Military Affairs, chapter 9—The Adjutant General).

RHODE ISLAND

Emergency Management

Director: Major General Robert T. Bray

Functions/Responsibilities: R.I. Gen. Laws § 30–15–5. The adjutant general, subject to the direction and control of the governor, shall be the executive head of the agency, and shall be responsible to the governor for carrying out the program for disaster preparedness of this state. The adjutant general shall coordinate the activities of all organizations for disasters within the state, and shall maintain liaison with and cooperate with disaster agencies and organizations of other states and of the federal government. The adjutant general also serves as Homeland Security Advisor. (Source: http://www.dhs.gov/xgovt/grants/states/rhodeisland.shtm)

Qualifications/Statutory Authority: R.I. Gen. Laws § 30–15–5. There is hereby created within the executive department, the Rhode Island emergency management agency (hereinafter in this chapter called the "agency"), to be headed by the adjutant general of the Rhode Island national guard who shall be appointed by and serve at the pleasure of, the governor, and who shall be in the unclassified service. (Source: http://law.justia.com/rhodeisland/codes/title30/30-15-5.html.)

Source: http://law.justia.com/rhodeisland/codes/title30/30-15-5.html (Title 30—Military Affairs and Defense, chapter 30-15—Emergency Management).

Adjutant General

Director: Major General Robert T. Bray

Functions/Responsibilities: R.I. Gen. Laws § 30–2–2. The adjutant general shall be chief of staff, commanding general of the Rhode Island army and air national guard, and paymaster general with the rank not to exceed that of major general. The adjutant general shall perform such duties as may be required by chapters 1–14 and any other law of the state, and such other duties as pertain to the adjutant general under the regulations and customs of the United States department of defense and the laws of Rhode Island. (Source: http://www.rilin.state.ri.us/Statutes/TITLE30/30-2/30-2-12.HTM.)

Qualifications: R.I. Gen. Laws § 30–2–13. No person shall be eligible to hold the office of adjutant general unless he or she holds or has held a commission of at least colonel in the armed forces of the United States, or in a reserve component thereof, and shall have served not less than five (5) years in one or more of the federal services, and shall meet the criteria for federal recognition in the rank to which he or she has been appointed as prescribed by the laws and regulations of the United States. (Source: http://www.rilin.state.ri.us/Statutes/TITLE30/30-2/30-2-13.HTM.)

Statutory Authority: R.I. Gen. Laws § 30–2–12. There shall be an adjutant general with rank not to exceed that of lieutenant general. The adjutant general shall be appointed by the governor and shall hold office for a term of four (4) years from the time of appointment and until his or her successor shall be appointed in his or her place and stead, provided, however, that this appointment may be revoked by the governor for cause or if the adjutant general shall have been found to be physically unfit for service, as provided by § 30–3–22.

Source: http://www.rilin.state.ri.us/Statutes/TITLE30/30-2/30-2-12.HTM (Title 30, chapter 30–2—Organization and Command of Military and Naval Forces).

SOUTH CAROLINA

Emergency Management Division

Director: Ronald C. Osborne

Functions/Responsibilities: S.C. Code § 25–1–420. The division must be administered by a director appointed by the Adjutant General, to serve at his pleasure, and such additional staff as may be employed or appointed by the Adjutant General. Functions include: (a) coordinating the efforts of all state, county, and municipal agencies and departments in developing a State Emergency Plan; (b) conducting a statewide preparedness program to assure the capability of state, county, and municipal governments to execute the State Emergency Plan; (c) establishing and maintaining a State Emergency Operations Center and providing support of the state emergency staff and work force; and (d) establishing an effective system for reporting, analyzing, displaying, and disseminating emergency information.

Qualifications/Statutory Authority: S.C. Code § 25–1–420. There is established within the office of the Adjutant General the South Carolina Emergency Management Division. The division must be administered by a director appointed by the Adjutant General, to serve at his pleasure, and such additional staff as may be employed or appointed by the Adjutant General.

Source: http://www.scstatehouse.net/cgi-bin/query.exe?first=DOC&querytext=%22Emergency %20Management%22&category=Code&conid=3217014&result_pos=0&keyval=423 (Title 25—Military, Civil Defense and Veterans Affairs, section 25-1-420—South Carolina Emergency Preparedness Division).

State Law Enforcement Division (SLED)

Director: Robert M. Stewart

Functions/Responsibilities: Exec. Order 2003–02 (January 2003) directs SLED to be the lead agency in the counterterrorism effort, including preparation against acts of terrorism in or affecting South Carolina and in the crisis management response to such acts. SLED is also directed to work closely with the Emergency Management Division and other government and private entities relevant to the homeland security mission. The Chief of SLED shall serve as the

Governor's representative to the United States Office of Homeland Security. (Source: http://www.sc gov/documents/CounterTerrorism/ExecutiveOrder2003-2.pdf.)

Qualifications: SLED Web site: The Chief of the State Law Enforcement Division (SLED) is the Chief Executive Officer of the agency. The Governor of South Carolina with the advice and consent of the Senate appoints the Chief of SLED. (Source: http://www.sled.sc.gov/ChiefOfSled. aspx?MenuID=ContactInformation.)

Statutory Authority: Exec. Order 2003–02 (January 2003). The South Carolina Law Enforcement Division (SLED) shall be the operational authority and lead agency in the counter-terrorism effort. The Chief of SLED shall create task forces or coordinating councils deemed appropriate to support this mission, and shall serve as the Governor's representative to the United States Office of Homeland Security.

Source: http://www.scgov/documents/CounterTerrorism/ExecutiveOrder2003-2.pdf.

Adjutant General

Director: Major General Stanhope S. Spears

Functions/Responsibilities: S.C. Code § 25–1–350. (1) Appoint the civilian employees of his department and he may remove any of them at his discretion; (2) keep rosters of all active, reserve and retired officers of the militia of the State, keep in his office all records and papers required to be kept and filed therein and submit to the Governor each year a printed annual report of the operations and conditions of the National Guard of South Carolina; (3) on the first day of July of each year, make a statement of the condition of the military fund, showing the amount thereof and setting forth in detail all receipts from whatsoever source and all expenditures of whatsoever nature and the unexpended balance thereof.

Qualifications: S.C. Code § 25–1–340. If the office of Adjutant General is vacated because of the death, resignation, or retirement of the Adjutant General prior to the normal expiration of his term of office, the Governor shall appoint an officer of the active South Carolina National Guard, who is at least the rank of lieutenant colonel, meets the eligibility requirements for a constitutional officer, and who has a minimum of fifteen years' active commissioned service in the South Carolina National Guard, to fill out the unexpired term of the former incumbent.

Statutory Authority: S.C. Code § 25–1–320. There shall be an Adjutant General elected by the qualified electors of this State at the same time and in the same manner and for the same term of office as other State officials. His rank shall be that of major-general. He shall hold office until his successor is elected and qualifies. He shall be ex officio chief of staff. He shall receive such annual salary as may be provided by the General Assembly.

Source: http://www.scstatehouse.net/cgi-bin/query.exe?first=DOC&querytext=Adjutant%20 General&category=Code&conid=3143743&result_pos=0&keyval=423 (Title 25, article 3— Military Department).

SOUTH DAKOTA

Office of Emergency Management

Director: Kristi Turman

Functions/Responsibilities: S.D. Codified Laws § 33–15–22. The secretary of public safety, subject to the direction and control of the Governor, is responsible for carrying out the program for emergency management of this state. He shall coordinate the activities of all organizations for emergency management within the state, and shall maintain liaison with and cooperation with emergency management agencies and organizations of other states and of the federal government. (Source: http://law.justia.com/southdakota/codes/33/33-15-22.html.) S.D. Codified Laws § 33–15–1. "Emergency management," the preparation for and the carrying out of all emergency functions, other than functions for which military forces are primarily responsible, to prevent, minimize, repair injury and damage resulting from disasters caused by enemy attack, sabotage, or other hostile action, fire, flood, snowstorm, windstorm, tornado, cyclone, drought, earthquake, or other natural causes and provide for the relief of distressed humans and livestock in areas where such conditions prevail whether affecting all or only a portion of the state. (Source: http://law.justia.com/southdakota/codes/33/33-15-1.html.)

Qualifications: S.D. Codified Laws § 1–33–4. Except as provided by § 1–33–10, the heads of the bureaus within the Department of Executive Management shall be appointed by the Governor and serve at his pleasure, and shall each have the title of commissioner. (Source: http://law.justia.com/southdakota/codes/1/1-33-4.html.)

Statutory Authority: S.D. Codified Laws § 33–15–2. To create a Division of Emergency Management, and to authorize the creation of local organizations for emergency management in the political subdivisions of the state.

Source: http://law.justia.com/southdakota/codes/33/33-15.html (Title 33— Military Affairs, chapter 15—Emergency Management).

Office of Homeland Security

Director: John Berheim

Functions/Responsibilities: South Dakota Homeland Security Mission Statement: We will lead the effort in keeping South Dakota free from any acts of terrorism. The Office of Homeland Security promotes its mission by coordinating an extensive information-sharing network between all levels of government and local officials; assisting all city, county and tribal governments with an ongoing assessment of their jurisdictions to determine their anti-terrorism needs; and managing anti-terrorism Homeland Security grants to assist city, county and tribal governments with the acquisition of resources needed to both prevent acts of terrorism and to respond and recover should one occur. (Source: http://www.state.sd.us/homeland/organization/organization_1.asp.)

Qualifications: S.D. Codified Laws § 1–33–4. Except as provided by § 1–33–10, the heads of the bureaus within the Department of Executive Management shall be appointed by the Governor and serve at his pleasure, and shall each have the title of commissioner. (Source: http://law.justia.com/southdakota/codes/1/1-33-4.html.)

Statutory Authority: Per the Homeland Security Director: There are no authorizing statutes or executive orders.

Adjutant General

Director: Major General Michael A. Gorman

Functions/Responsibilities: S.D. Codified Laws § 33–1–12. The adjutant general shall have general supervision and control of the Department of Military and Veterans Affairs subject to the orders and instructions of the Governor, and he shall have such staff assistants in the several divisions named in this chapter as he shall recommend and the Governor shall deem necessary for economical administration. S.D. Codified Laws § 33–1–6. The adjutant general shall distribute all orders from the Governor. The adjutant general is the organ of all written communication from the National Guard to the Governor and shall attend the Governor if required at review of the National Guard, or if ordered in the performance of military duty. The adjutant general shall present to the Governor all recommendations with reference to the Department of Military and Veterans Affairs and shall obey and issue orders given by the Governor in relation to the department and in all other military matters. (Source: http://legis. state.sd.us/statutes/DisplayStatute.aspx?Type=Statute&Statute=33-1-6.)

Qualifications: S.D. Codified Laws § 33–1–2. One adjutant general, in the grade of major general, shall be appointed and shall serve as provided by § 1–32–3. At the time of appointment, the adjutant general shall be a federally recognized commissioned officer of the South Dakota National Guard, with not less than ten years military service in the armed forces of this state or of the United States. The officer appointed to the position of adjutant general shall meet all of the requirements of the officer's respective service to be appointed and receive federal recognition as a general officer in that service, including any waivers that may be authorized and granted or delegated by the secretaries of the Army or Air Force, as appropriate. (Source: http://legis.state. sd.us/statutes/DisplayStatute.aspx?Type=Statute&Statute=33-1-2.)

Statutory Authority: S.D. Codified Laws § 33–1–2 (2007). One adjutant general, in the grade of major general, shall be appointed and shall serve as provided by § 1–32–3.

Source: http://law.justia.com/southdakota/codes/33/33-1.html (Title 33, chapter 1—State Department of Military and Veterans Affairs).

TENNESSEE

Emergency Management Agency

Director: Jim Bassham

Functions/Responsibilities: Tenn. Code Ann. § 58–2–106. The agency is responsible for maintaining a comprehensive statewide program of emergency management. The agency is responsible for coordination with efforts of the federal government with other departments and agencies of state government, county governments, municipal governments and school boards, and private agencies that have a role in emergency management. Tenn. Code Ann. § 58–2–104. The director, subject to the direction and control of the governor, acting through the adjutant general, shall be the executive head of the agency and shall be responsible to the governor for carrying out the program for TEMA for the state of Tennessee. The director shall coordinate the activities of all organizations for the agency within the state and shall maintain liaison with and cooperate with emergency management agencies and organizations of other states and of the federal government. For normal day-to-day administrative functions, the director shall report to the adjutant general. During emergency conditions, the agency and director shall report to the governor or the governor's designee. General coordination with the adjutant general shall be maintained. The department of the military shall become a resource for the state as with all other departments and agencies; further, the director shall make recommendations to the governor for the use of the national guard and other state resources as disaster conditions mandate, including, but not limited to, the assistance of local and private agencies. The director shall coordinate with the governor's office on the activation or the potential activation of any mutual aid agreement or compact.

Qualifications/Statutory Authority: Tenn. Code Ann. § 58–2–104. The governor is hereby authorized and directed to create a state agency to be known as the "Tennessee emergency management agency" (TEMA) under the adjutant general for day-to-day administrative purposes and, upon the recommendation of the adjutant general, to appoint a director of the TEMA, who shall be the administrator thereof.

Source: http://michie.lexisnexis.com/tennessee/lpext.dll?f=templates&fn=main-h.htm&cp=tn code (Title 58—Military Affairs, Emergencies and Civil Defense, chapter 2—Disasters, Emergencies and Civil Defense).

Department of Public Safety, Office of Homeland Security

Director: David Mitchell

Functions/Responsibilities: Exec. Order. No. 8 (April 2003). The responsibilities of the Council are to work with the Office of Homeland Security in planning and directing statewide homeland security activities and to interact with federal and statewide homeland security activities and to interact with federal and local officials in promoting homeland security. The Council will be chaired by the Director of Homeland Security. The Office of Homeland Security is hereby designated as the office having primary responsibility and authority for directing the state's homeland security activities, including but not limited to the planning, coordination and implementation of all homeland security prevention, protection and response operations. This responsibility shall specifically include the duty to design, develop and implement a comprehensive, coordinated strategy to secure the state of Tennessee from terrorist threats and attacks.

Qualifications/Statutory Authority: Exec. Order. No. 8 (April 2003). There is hereby constituted the Office of Homeland Security to be operated under the authority and supervision of the Director of Homeland Security. The Director of the Office of Homeland Security shall be appointed by the Governor, shall serve as a member of the Governor's Cabinet and shall report directly to the Governor on all homeland security matters.

Source: http://www.state.tn.us/sos/pub/execorders/exec-orders-bred8.pdf.

Adjutant General

Director: Major General Gus L. Hargett

Functions/Responsibilities: Tenn. Code Ann. § 58–1–114. The adjutant general shall be the executive head of the military department and commanding general of the military forces of the state. Tenn. Code Ann. § 58–1–116. (a) The adjutant general shall be chief of staff to the governor and subordinate only to the governor in matters pertaining to the military department and the military and naval affairs of the state. (b) It shall be the duty of the adjutant general to direct the planning and employment of the military forces of the state in carrying out their state military mission; to establish unified command of state forces whenever they shall be jointly engaged; to coordinate the military and naval affairs with the civil defense of the state.

Qualifications: Tenn. Code Ann. § 58–1–115. The adjutant general shall be appointed by the governor for a term concurrent with the term of the governor who appointed the adjutant general, and shall serve as such at the pleasure of the governor. The adjutant general shall have such rank as may be conferred by the governor, but in no event shall such rank be higher than that of lieutenant general.

Statutory Authority: Tenn. Code Ann. § 58–1–110. The military staff of the governor shall consist of the adjutant general, who shall be ex officio chief of staff, the national guard officers assigned to the state headquarters, and such aides-de-camp as the governor shall deem necessary, all of whom shall be federally recognized, commissioned officers detailed by the governor from the national guard.

Source: http://michie.lexisnexis.com/tennessee/lpext.dll?f=templates&fn=main-h.htm&cp=tn code (Title 58, chapter 1—Military Forces).

TEXAS

Emergency Management

Director: Jack Colley

Functions/Responsibilities: Tex. Gov't. Code § 418.041. The division is managed by a director appointed by the governor. The director serves at the pleasure of the governor. (Source: http://tlo2.tlc.state.tx.us/statutes/gv.toc.htm.) Tex. Gov't. Code § 418.042. The division shall prepare and keep current a comprehensive state emergency management plan. In preparing and revising the state emergency management plan, the division shall seek the advice and assistance of local government, business, labor, industry, agriculture, civic organizations, volunteer organizations, and community leaders. All or part of the state emergency management plan may be incorporated into regulations of the division or executive orders that have the force and effect of law.

Qualifications/Statutory Authority: Tex. Gov't. Code § 418.041. The division of emergency management is a division of the office of the governor. The division is managed by a director appointed by the governor.

Source: http://law.justia.com/texas/codes/gv/004.00.000418.00.html (chapter 418—Emergency Management).

Office of Homeland Security

Director: Steven McCraw

Functions/Responsibilities: Governor's Press Release (August 2004): Steven McCraw, Director of Homeland Security, will be responsible for directing day-to-day homeland security efforts among local, state and federal agencies. (Source: http://www.governor.state.tx.us/divisions/press/pressreleases/ PressRelease.2004-08-02.0647/view.) Tex. Gov't. Code § 421.024. The council shall advise the governor on: (1) the development and coordination of a statewide critical infrastructure protection strategy; (2) the implementation of the governor's homeland security strategy by state and local agencies and provide specific suggestions for helping those agencies implement the strategy; and (3) other matters related to the planning, development, coordination, and implementation of initiatives to promote the governor's homeland security strategy.

Qualifications: Tex. Gov't. Code § 421.043. (a) To be eligible for appointment as a member of a permanent special advisory committee created under this subchapter, a person must demonstrate experience in the sector that the person is under consideration to represent and be directly involved in related policies, programs, or funding activities that are relevant to homeland security or infrastructure protection. (b) Each member of a permanent special advisory committee created under this subchapter serves at the will of the governor.

Statutory Authority: Tex. Gov't. Code § 421.023. (a) The council is an advisory entity administered by the office of the governor. (b) The governor may adopt rules as necessary for the operation of the council. (c) The governor shall designate the presiding officer of the council.

Source: http://law.justia.com/texas/codes/gv/004.00.000421.00.html (chapter 421—Homeland Security).

Adjutant General

Director: Lieutenant General Charles G. Rodriguez

Functions/Responsibilities: Tex. Gov't. Code § 431.022. The adjutant general is the head of the adjutant general's department and controls the military department of the state. The adjutant general is subordinate only to the governor in matters pertaining to the military department of the state and the state military forces. Tex. Gov't. Code § 431.029. The adjutant general shall: (1) perform duties that the governor assigns relating to the military affairs of the state and conduct the business of the department as the governor directs; (2) perform for the state as near as practicable the duties that pertain to the chiefs of staff of the army and air force and the secretaries of the military services, under regulations and customs of the United States armed forces; (3) control and supervise the transportation of troops, munitions of war, military equipment and property, and stores in the state.

Qualifications: Tex. Gov't. Code § 431.022. The adjutant general is appointed by the governor, with the advice and consent of the senate if in session, to a term expiring February 1 of each odd-numbered year. To be qualified for appointment as adjutant general a person must: (1) when appointed be serving as a federally recognized officer of not less than field grade in the Texas National Guard; (2) have previously served on active duty or active duty for training with the army, air force, or marines; and (3) have completed at least 10 years' service as a federally recognized reserve or active duty commissioned officer with an active unit of the United States armed forces, the National Guard, or the Texas National Guard, including at least five years with the Texas National Guard. The appointment of the adjutant general shall be made without regard to the race, color, sex, religion, or national origin of the appointee.

Statutory Authority: Tex. Gov't. Code § 431.003. The governor's military staff consists of: the adjutant general.

Source: http://law.justia.com/texas/codes/gv/004.00.000431.00.html (section 431—State Militia).

UTAH

Department of Public Safety, Division of Homeland Security

Director: Michael Kuehn

Functions/Responsibilities: Utah Code § 53–2–103. The division shall be administered by a director appointed by the commissioner with the approval of the governor. Utah Code § 53–2–104. The division shall: (a) respond to the policies of the governor and the Legislature; (b) perform functions relating to emergency services and homeland security matters as directed by the commissioner; and (c) prepare, implement, and maintain programs and plans. The division coordinates emergency management efforts (preparedness, recovery, response, and mitigation) between federal, state and local governments. (Source: http://homelandsecurity.utah.gov/home/aboutus)

Qualifications/Statutory Authority: Utah Code § 53–2–103. (1) There is created within the department the Division of Homeland Security. (2) The division shall be administered by a director appointed by the commissioner with the approval of the governor. (3) The director is the executive and administrative head of the division and shall be experienced in administration and possess additional qualifications as determined by the commissioner and as provided by law.

Source: http://www.livepublish.le.state.ut.us/lpBin22/lpext.dll?f=templates&fn=main-j.htm&2.0 (Title 53—Public Safety Code, chapter 2—Emergency Management, sections 103–104—Division of Homeland Security).

Adjutant General

Director: Major General Brian L. Tarbet

Functions/Responsibilities: Utah Code § 39–1–12. (1) There shall be one adjutant general appointed by the governor. The adjutant general is chief of staff and holds office for a term of six years, unless terminated by resignation, disability, or for cause as determined by a military court or court-martial. (2) He shall perform duties as are imposed by the laws of this state and the United States, and by the regulations of the Department of Defense of the United States. However, if any duties imposed by the statutes of this state at any later time conflict with those imposed by the laws of the United States, the duties imposed by the statutes of this state, as far as they conflict, are abrogated. He has, under the direction of the State Armory Board, supervision and charge of all the armories, warehouses, maintenance and repair shops, hangars, small-arms, artillery and aircraft ranges, campsites, concentration areas, lands, training facilities, and military reservations necessary to the military functions of this state. He is responsible for the protection and safety thereof and shall make rules for the maintenance of order, for the enforcement of rules as may be ordered for the operation and the repair, care, and preservation of the facilities and installations belonging to or leased by the state. He may make further improvement as the good of the service requires.

Qualifications: Utah Code § 39–1–12. The person appointed to the office shall be a citizen of Utah and meet the requirements provided in Title 32, United States Code. He shall be a federally recognized commissioned officer of the National Guard of the United States with no fewer than ten years commissioned service in the Utah National Guard. Active service in the armed forces of the United States may be included in this requirement, if the officer was a member of the Utah National Guard when he entered that service. An officer is no longer eligible to hold the office of adjutant general after becoming 64 years of age.

Statutory Authority: Utah Code § 39–1–12. There shall be one adjutant general appointed by the governor. The adjutant general is chief of staff and holds office for a term of six years, unless terminated by resignation, disability, or for cause as determined by a military court or court-martial.

Source: http://www.livepublish.le.state.ut.us/lpBin22/lpext.dll?f=templates&fn=main-j.htm&2.0 (Title 39—Militia and Armories, chapter 1—State Militia, section 12—Adjutant General).

VERMONT

Emergency Management

Director: Barbara Farr

Functions/Responsibilities: 20 Vt. Stat. § 3. There shall be a director of Vermont emergency management who shall be in immediate charge of the division. The director shall: (1) Coordinate the activities of all emergency management organizations within the state. (2) Maintain liaison and cooperation with emergency management agencies and organizations of the federal government, other states, and Canada. (3) Perform additional duties and responsibilities required pursuant to this chapter and prescribed by the governor. The commissioner, subject to the approval of the governor, shall delegate to the several departments and agencies of the state government appropriate emergency management responsibilities, and review and coordinate the emergency management activities of the departments and agencies with each other and with the activities of the districts and neighboring states, the neighboring Canadian province of Quebec, and the federal government.

Qualifications/Statutory Authority: 20 Vt. Stat. § 3. There shall be a director of Vermont emergency management who shall be in immediate charge of the division. The director shall be appointed by the commissioner, with the approval of the governor. The director shall serve at the pleasure of the commissioner and shall hold no other state office.

Source: http://michie.lexisnexis.com/vermont/lpext.dll?f=templates&fn=main-h.htm&cp= (Title 20—Internal Security and Public Safety, chapter 1; Civil Defense, section 3—Vermont Emergency Management Division).

Department of Public Safety, Homeland Security Unit

Director: Kerry Sleeper

Functions/Responsibilities: Exec. Order 20–29 (February 2003). The mission of the Governor's Homeland Security Advisory Council (herein, "the Council") shall be to assess the State's overall homeland security preparedness, policies, and communications and to advise on strategies to improve our current system. The Council shall carefully consider the interdependencies between federal, state, and local governments, the Vermont National Guard, first responders, law enforcement, emergency managers, public health officials, and private community organizations and the programs these groups administer and advise on strategies to strengthen and bolster those relationships. The Council shall be mindful of available financial resources and advise on strategies that are consistent with existing fiscal realities.

Qualifications: Exec. Order 20–29 (February 2003). The Governor shall appoint the Chair of the Council.

Statutory Authority: Exec. Order 20–29 (February 2003). The Vermont Terrorism Task Force is renamed the "Governor's Homeland Security Advisory Council" and redefined and reconstituted as set forth below.

Source: http://www.leg.state.vt.us/statutes/fullsection.cfm?Title=03APPENDIX&Chapter= 020&Section=00029.

Adjutant General

Director: Major General Michael D. Dubie

Functions/Responsibilities: 20 Vt. Stat. § 361. The military department, created by section 212 of Title 3, shall be administered by the adjutant general and shall include the national guard and all military components of the state. 20 Vt. Stat. § 1. The Adjutant General of the State of Vermont is authorized and directed to take all appropriate steps to organize and maintain the formation of the Vermont State Guard Cadre in an inactive status without pay, to serve for an indefinite period and until relieved by proper authority.

Qualifications: 20 Vt. Stat. § 363. The general assembly shall biennially elect an adjutant and inspector general, who shall also be quartermaster general with the rank of a major general.

Statutory Authority: 20 Vt. Stat. § 1. The Adjutant General of the State of Vermont is authorized and directed to take all appropriate steps to organize and maintain the formation of the Vermont State Guard Cadre in an inactive status without pay, to serve for an indefinite period and until relieved by proper authority.

Source: http://michie.lexisnexis.com/vermont/lpext.dll?f=templates&fn=main-h.htm&cp= (Title 20, chapter 21, sections 361—National Guard and section 363—Officers Generally, and chapter 1, section 1—Purpose and Policy).

VIRGINIA

Emergency Management

State Coordinator: Michael Cline

Functions/Responsibilities: Va. Code § 44–146–17. The Governor shall be Director of Emergency Management. He shall take such action from time to time as is necessary for the adequate promotion and coordination of state and local emergency services activities relating to the safety and welfare of the Commonwealth in time of natural or man-made disasters. (Source: http://leg1.state.va.us/cgi-bin/legp504.exe?000+cod+44-146.17.) Va. Code § 44–146.18. In coordination with political subdivisions and state agencies, ensure that the Commonwealth has up-to-date assessments and preparedness plans to prevent, respond to and recover from all disasters including acts of terrorism; Conduct a statewide emergency management assessment in cooperation with political subdivisions, private industry and other public and private entities deemed vital to preparedness, public safety and security. The assessment shall include a review of emergency response plans, which include the variety of hazards, natural and man-made. The assessment shall be updated annually; Submit to the Governor and to the General Assembly, no later than the first day of each regular session of the General Assembly, an annual executive summary and report on the status of emergency management response plans throughout the Commonwealth and other measures taken or recommended to prevent, respond to and recover from disasters, including acts of terrorism. (Source: http://leg1.state.va.us/cgi-bin/legp504.exe?000+cod+44-146.18.)

Qualifications: Va. Code § 44–147.17. The Governor has the power to appoint a State Coordinator of Emergency Management and authorize the appointment or employment of other personnel as is necessary to carry out the provisions of this chapter, and to remove, in his discretion, any and all persons serving hereunder. (Source: http://leg1.state.va.us/cgi-bin/legp504.exe?000+cod+44-146.17.)

Statutory Authority: Va. Code § 44–146.14. To create a State Department of Emergency Management, and to authorize the creation of local organizations for emergency management in the political subdivisions of the Commonwealth.

Source: http://leg1.state.va.us/cgi-bin/legp504.exe?000+cod+2.2-305 (Title 44—Military and Emergency Laws, chapter 3.2—Emergency Services and Disaster Law).

Office of Commonwealth Preparedness

Director: Robert P. Crouch Jr.

Functions/Responsibilities: Va. Code § 2.2–304. The purpose of the Office is to work with and through others, including federal, state, and local officials, as well as the private sector, to develop a seamless, coordinated security and preparedness strategy and implementation plan. (Source: http://leg1.state.va.us/cgi-bin/legp504.exe?000+cod+2.2-304.) Va. Code § 2.2–305. 1. Provide oversight, coordination, and review of all disaster, emergency management, and terrorism management plans for the state and its agencies. Work with federal officials to obtain additional federal resources and coordinate policy development and information exchange. Coordinate working relationships between state agencies and the Governor's Cabinet. (Source: http://leg1.state.va.us/cgi-bin/legp504.exe?000+cod+2.2-305._)

Qualifications: Va. Code § 2.2–304. The Office shall consist of an Assistant to the Governor for Commonwealth Preparedness (the Assistant to the Governor) who shall be appointed by the Governor. The appointment shall be subject to confirmation by the General Assembly as provided in § 2.2–106. The Assistant to the Governor shall, by reason of professional background have knowledge of military, law enforcement, public safety, or emergency management and preparedness issues, in addition to familiarity with the structure and operations of the federal government and of the Commonwealth. (Source: http://leg1.state.va.us/cgi-bin/legp504.exe? 000+cod+2.2-304.)

Statutory Authority: Va. Code § 2.2–304. There is created in the Office of the Governor, the Office of Commonwealth Preparedness (the Office).

Source: http://leg1.state.va.us/cgi-bin/legp504.exe?000+cod+2.2-304 (Title 2.2—Administration of Government, chapter 3.1—Office of Commonwealth Preparedness).

Adjutant General

Director: Major General Robert B. Newman, Jr.

Functions/Responsibilities: Va. Code § 44–13. As head of the Department of Military Affairs, the Adjutant General shall have command of all of the militia of the Commonwealth, subject to the orders of the Governor as Commander in Chief, and shall distribute all orders from the Governor pertaining to the military service and shall perform all duties imposed upon him or that Department by this title in the manner prescribed by law. (Source: http://leg1.state.va.us/cgi-bin/legp504.exe?000+cod+44-13.) Va. Code § 44–11. The Adjutant General shall be in direct charge of the Department of Military Affairs and shall be responsible to the Governor and commander in chief for the proper performance of his duties. All the powers conferred and the duties imposed by law upon the Adjutant General shall be exercised or performed by him under the direction and control of the Governor. (Source: http://leg1.state.va.us/cgi-bin/legp504.exe? 000+cod+44-11.)

Qualifications/Statutory Authority: Va. Code § 44–11. There is hereby created the Department of Military Affairs to which is transferred all of the functions, powers and duties of the former Division of Military Affairs. The Governor shall appoint an Adjutant General with the rank of brigadier general, major general or lieutenant general as the Governor may prescribe, subject to confirmation by the General Assembly if in session, and if not in session, then at its next succeeding session. The Adjutant General shall not hold the rank of lieutenant general unless such rank is federally recognized. No person shall be appointed Adjutant General who shall not have had at least ten years' commissioned service in the Virginia National Guard in at least field grade.

Source: http://leg1.state.va.us/cgi-bin/legp504.exe?000+cod+44-11 (Title 44, chapter 1—Military Laws of Virginia).

WASHINGTON

Military Department, Emergency Management Division

Director: Jim Mullen

Functions/Responsibilities: Wash. Rev. Code § 38.52.030. The director shall coordinate the activities of all organizations for emergency management within the state, and shall maintain liaison with and cooperate with emergency management agencies and organizations of other states and of the federal government, and shall have such additional authority, duties, and responsibilities authorized by this chapter, as may be prescribed by the governor. The director shall develop and maintain a comprehensive, all-hazard emergency plan for the state which shall include an analysis of the natural, technological, or human caused hazards which could affect the state of Washington, and shall include the procedures to be used during emergencies for coordinating local resources, as necessary, and the resources of all state agencies, departments, commissions, and boards. The comprehensive emergency management plan shall direct the department in times of state emergency to administer and manage the state's emergency operations center. The director shall make such studies and surveys of the industries, resources, and facilities in this state as may be necessary to ascertain the capabilities of the state for emergency management, and shall plan for the most efficient emergency use thereof. (Source: http://search.leg.wa.gov/pub/textsearch/ViewRoot.asp?Action=Html&Item=0&X= 910074535&p=1.)

Qualifications: Wash. Rev. Code § 38–52–010. "Director" means the adjutant general. (Source: http://apps.leg.wa.gov/RCW/default.aspx?cite=38.52.010.) Wash. Rev. Code § 38.12.030. Whenever a vacancy has occurred, or is about to occur in the office of the adjutant general, the governor shall order to active service for that position from the active list of the Washington army national guard or Washington air national guard an officer not below the rank of a field grade officer who has had at least ten years service as an officer on the active list of the Washington army national guard or the Washington air national guard during the fifteen years next prior to such detail. The officer so detailed shall during the continuance of his or her service as the adjutant general hold the rank of a general officer. (Source: http://apps.leg.wa.gov/RCW/ default.aspx?cite=38.12.030.)

Statutory Authority: Wash. Rev. Code § 38.52.020. Because of the existing and increasing possibility of the occurrence of disasters of unprecedented size and destructiveness as defined in RCW 38.52.010(6), and in order to insure that preparations of this state will be adequate to deal with such disasters, to insure the administration of state and federal programs providing disaster relief to individuals, and further to insure adequate support for search and rescue operations, and generally to protect the public peace, health, and safety, and to preserve the lives and property of the people of the state, it is hereby found and declared to be necessary: To provide for emergency management by the state, and to authorize the creation of local organizations for emergency management in the political subdivisions of the state.

Source: http://apps.leg.wa.gov/RCW/default.aspx?cite=38.52.020 (Title 38—Militia and Military Affairs, chapter 38.52—Emergency Management).

Adjutant General

Director: Major General Timothy J. Lowenberg

Functions/Responsibilities: Wash. Rev. Code § 38.08.020. The militia of the state not in the service of the United States shall be governed and its affairs administered pursuant to law, by the governor, as commander-in-chief, through the adjutant general's department, of which the adjutant general shall be the executive head. (Source: http://apps.leg.wa.gov/RCW/default.aspx?

cite=38.08.020.) Wash. Rev. Code § 38.12.020. Keep rosters of all active, reserve, and retired officers of the militia, and all other records, and papers required to be kept and filed therein, and shall submit to the governor such reports of the operations and conditions of the organized militia as the governor may require. Attend to the care, preservation, safekeeping, and repairing of the arms, ordnance, accoutrements, equipment, and all other military property belonging to the state, or issued to the state by the United States for military purposes, and keep accurate accounts thereof. Any property of the state military department which, after proper inspection, is found unsuitable or no longer needed for use of the state military forces, shall be disposed of in such manner as the governor shall direct and the proceeds thereof used for replacements in kind or by other needed authorized military supplies, and the adjutant general may execute the necessary instruments of conveyance to effect such sale or disposal. Major General Lowenberg is also Homeland Security Advisor to the governor. (Source: http://apps.leg.wa.gov/RCW/default.aspx?cite=38.12.020.)

Qualifications/Statutory Authority: Wash. Rev. Code § 38.12.030. Whenever a vacancy has occurred, or is about to occur in the office of the adjutant general, the governor shall order to active service for that position from the active list of the Washington army national guard or Washington air national guard an officer not below the rank of a field grade officer who has had at least ten years service as an officer on the active list of the Washington army national guard or the Washington air national guard during the fifteen years next prior to such detail.

Source: http://apps.leg.wa.gov/RCW/default.aspx?cite=38.12.030 (Title 38, chapter 38.12—Militia Officers).

WEST VIRGINIA

Division of Homeland Security and Emergency Management

Director: Jimmy Gianato

Functions/Responsibilities: W. Va. Code § 15–5–3. The Director, subject to the direction and control of the Governor through the Secretary of the Department of Military Affairs and Public Safety, shall be executive head of the Division of Homeland Security and Emergency Management and shall be responsible to the Governor and the Secretary of the Department of Military Affairs and Public Safety for carrying out the program for homeland security and emergency management in this state. The Director in consultation with the Secretary of the Department of Military Affairs and Public Safety shall coordinate the activities of all organizations for homeland security and emergency management within the state and maintain liaison with and cooperate with homeland security, emergency management and other emergency service and civil defense agencies and organizations of other states and of the federal government, and shall have additional authority, duties and responsibilities authorized by this article as may be prescribed by the Governor or the Secretary of the Department of Military Affairs and Public Safety.

Qualifications: W. Va. Code § 15–5–3. A Director of the Division of Homeland Security and Emergency Management shall be appointed by the Governor, by and with the advice and consent of the Senate. The Governor shall consider applicants for Director who at a minimum: (1) Have at least five years managerial or strategic planning experience; (2) are knowledgeable in matters relating to public safety, homeland security, emergency management and emergency response;

and (3) have at a minimum, a federally issued secret level security clearance or have submitted to or will submit to a security clearance investigation for the purpose of obtaining, at a minimum, a federally issued secret level security clearance.

Statutory Authority: W. Va. Code § 15–5–3. Division of Homeland Security and Emergency Management created. The Office of Emergency Services is continued as the Division of Homeland Security and Emergency Management within the Department of Military Affairs and Public Safety.

Source: http://www.legis.state.wv.us/WVCODE/15/WVC%2015%20%20-%20%205%20%20-%20%20%203%20%20.htm (chapter 15—Public Safety, section 15–3–3—Division of Homeland Security and Emergency Management).

Adjutant General

Director: Major General Allen E. Tackett

Functions/Responsibilities: W. Va. Code § 15–1A–1. The adjutant general's department shall be a part of the executive branch of the government charged with the organization, administration, operation and training, supply and discipline of the military forces of the state. W. Va. Code § 15–1A–3. The adjutant general shall be chief of staff to the governor and commanding general of the organized militia. He shall direct the planning and employment of the military forces of the state in carrying out their state mission, establish unified command of state forces whenever jointly engaged, coordinate the military affairs with the civil defense of the state and organize and coordinate the activities of all civil agencies including local and state police in event of declaration of a limited emergency by the governor pursuant to article one-d of this chapter. In time of emergency or disaster, the adjutant general shall coordinate his activities with those of the office of emergency services provided for by article five of this chapter. He shall be custodian of all military records of the state and shall keep the same indexed and available for ready reference. He shall keep an itemized account of all moneys received and dispensed from all sources and shall make an annual report to the governor on the condition of the organized militia, receipts and expenditures and such other matters relating to the military forces of the state and the adjutant general's department as he shall deem expedient. (b) The adjutant general shall be responsible for the organization, administration, training and supply of the organized militia.

Qualifications/Statutory Authority: W. Va. Code § 15–1A–2. The adjutant general shall be appointed by the governor, by and with the advice and consent of the Senate, for a term of four years. He or she shall have the rank of major general, or such other rank as is recognized by federal authority. No person may be appointed adjutant general unless he or she has had at least six years' commissioned service and attained field grade or higher rank in the organized militia of this or some other state or in the armed forces of the United States, or in all combined. The governor shall require the adjutant general to furnish bond as required by law, which bond shall be filed with the auditor of the state.

Source: http://www.legis.state.wv.us/WVCODE/15/masterfrmFrm.htm (chapter 15, article 1A—Adjutant General).

WISCONSIN

Department of Military Affairs, Division of Emergency Management

Director: Brigadier General Donald Dunbar

Functions/Responsibilities: Wisc. Stat. § 166.03. (2)(a) The adjutant general shall: 1. Subject to approval by the governor, develop and promulgate a state plan of emergency management for the security of persons and property which shall be mandatory during a state of emergency. 2. Prescribe and carry out statewide training programs and exercises to develop emergency management proficiency, disseminate information including warnings of enemy action, serve as the principal assistant to the governor in the direction of emergency management activities and coordinate emergency management programs between counties. 3. Furnish guidance and develop and promulgate standards for emergency management programs for counties, cities, villages, and towns, and prescribe nomenclature for all levels of emergency management.

Qualifications: Wisc. Stat. § 15.31. A person must meet all of the following requirements to be appointed as the adjutant general: Hold the federally recognized minimum rank of full colonel. Except for those qualified under sub.(4), be a current participating member of one of the following components: (a) The Wisconsin army national guard. (b) The army national guard of the United States. (c) The U.S. army reserve. (d) The Wisconsin air national guard. (e) The air national guard of the United States. (f) The U.S. air force reserve. Be fully qualified to receive federal recognition at the minimum rank of brigadier general and have successfully completed a war college course or the military equivalent acceptable to the appropriate service. If the applicant is already a federally recognized general officer, meet all of the following conditions: (a) Be retired from active drilling status within the preceding 2 years. (b) The basis of the applicant's retired status was service with one of the service components noted in sub. (c) Be 62 years of age or less. (d) Continue to be eligible for federal recognition as a major general.

Statutory Authority: Wisc. Stat. § 15.313. There is created in the department of military affairs a division of emergency management. The administrator of this division shall be nominated by the governor and with the advice and consent of the senate appointed, to serve at the pleasure of the governor.

Source: http://nxt.legis.state.wi.us/nxt/gateway.dll?f=templates&fn=default.htm&vid=WI:Default&d=stats&jd=top (chapter 166—Emergency Management and chapter 15—Structure of the Executive Branch).

Homeland Security Council

State Homeland Security Advisor: Brigadier General Donald Dunbar

Functions/Responsibilities: Exec. Order # 7 (March 2003). Provide that the Council shall have the following missions: To advise the Governor and coordinate the efforts of state and local officials with regard to prevention of, and response to, threats to the homeland security of Wisconsin, addressing the coordination among state agencies that complement local initiatives and support the national homeland security principles of detection, deterrence, preparedness, response and recovery. Require the Council to submit periodic reports to the Governor with a final report due to the Governor, pursuant to the Governor's request.

Qualifications/Statutory Authority: Exec. Order # 7 (March 2003). Create the Governor's Homeland Security Council (hereinafter the "Council"); and Provide that the Council shall consist of seven (7) members appointed by the Governor to serve at the pleasure of the Governor; and Provide that the Governor shall designate one (1) member of the Council as chair to serve in that capacity at the pleasure of the Governor.

Source: http://www.wisgov.state.wi.us/journal_media_detail.asp?locid=19&prid=100.

Adjutant General

Director: Brigadier General Donald Dunbar

Functions/Responsibilities: Wisc. Stat. § 21.025. The adjutant general may establish a plan for organizing a military force to be known as the Wisconsin state defense force. The adjutant general may organize the Wisconsin state defense force under the plan if all or part of the national guard is called into the service of the United States. It shall be distinct from the national guard, uniformed, and composed of officers, commissioned or assigned, and of enlisted personnel who volunteer for service. Wisc. Stat. § 21.19. The adjutant general shall be chief of staff to the governor. The adjutant general shall have the custody of all property, military records, correspondence and other documents relating to the national guard and any other military forces organized under the laws of this state. The adjutant general may appoint an assistant quartermaster general to issue and account for state property. The adjutant general shall be the medium of military correspondence with the governor and perform all other duties pertaining to the office or prescribed by law, including the preparation and submission to the governor of reports under s. 15.04 (1) (d). The adjutant general shall administer, with the approval of the governor, state-federal cooperative funding agreements.

Qualifications: Wisc. Stat. § 15.31. A person must meet all of the following requirements to be appointed as the adjutant general: Hold the federally recognized minimum rank of full colonel. Except for those qualified under sub. (4), be a current participating member of one of the following components: (a) The Wisconsin army national guard. (b) The army national guard of the United States. (c) The U.S. army reserve. (d) The Wisconsin air national guard. (e) The air national guard of the United States. (f) The U.S. air force reserve. Be fully qualified to receive federal recognition at the minimum rank of brigadier general and have successfully completed a war college course or the military equivalent acceptable to the appropriate service. If the applicant is already a federally recognized general officer, meet all of the following conditions: (a) Be retired from active drilling status within the preceding 2 years. (b) The basis of the applicant's retired status was service with one of the service components noted in sub. (2). (c) Be 62 years of age or less. (d) Continue to be eligible for federal recognition as a major general.

Statutory Authority: Wisc. Stat. § 15.31. There is created a department of military affairs under the direction and supervision of the adjutant general who shall be appointed by the governor for a 5-year term.

Source: http://nxt.legis.state.wi.us/nxt/gateway.dll?f=templates&fn=default.htm&vid=WI:Defa ult&d=stats&jd=top (chapter 21—Department of Military Affairs, and chapter 15—Structure of the Executive Branch).

WYOMING

Homeland Security

Director: Joe Moore

Functions/Responsibilities: Wyo. Stat. § 19–13–104. Supervise the Wyoming office of homeland security; Provide technical assistance to public safety agencies in the area of homeland security; Coordinate with the federal department of homeland security. Wyo. Stat. § 19–13–105. The Director is the administrative head of the Wyoming office of homeland security. In addition to the duties described in W.S. 19–13–104 the director: Shall be responsible to the governor for the implementation of the state program for homeland security for Wyoming; Shall assist the local authorities and organizations in the planning and development of local homeland security plans and programs; Shall coordinate the activities of all organizations for homeland security within the state, including all state departments; Shall maintain liaison with and cooperate with homeland security agencies and programs of other states and of the federal government.

Qualifications/Statutory Authority: Wyo. Stat. § 19–13–104. The position of the director, office of homeland security is created in the governor's office and shall be appointed by the governor. He shall be responsible to the governor and may be removed by the governor as provided in W.S. 9–1–202.

Source: http://legisweb.state.wy.us/statutes/statutes.aspx?file=titles/Title19/Title19.htm (Title 19—Defense Forces and Affairs, chapter 13—Wyoming Office of Homeland Security).

Adjutant General

Director: Major General Edward L. Wright

Functions/Responsibilities: Wyo. Stat. § 19–7–103. The adjutant general of Wyoming shall have powers and duties and be paid a salary as follows: He is in control of the military department of Wyoming and subordinate only to the governor in matters pertaining thereto. He acts as the governor's designee with respect to personnel matters, including enlistments, promotions, demotions and discharges. He shall issue and transmit all orders of the commander in chief and make returns and reports as the secretary of defense may direct; He shall keep a record of all officers commissioned by the governor and all orders, rules and regulations prescribed by the secretary of defense and the several agencies of the department of defense for the national guard; He shall audit all claims and accounts against the military department not otherwise provided by law and shall have charge and carefully preserve the colors, flags and military trophies of war belonging to the state which shall not be loaned or removed from their prescribed place of deposit except by order of the adjutant general; He shall control all armories that are owned, erected, purchased, leased or provided by the state.

Qualifications: Wyo. Stat. § 19–7–103. The governor shall appoint a qualified adjutant general to be chief of staff. He shall have the rank the governor designates and may be removed by the governor as provided in W.S. 9–1–202. A prospective appointee shall possess the following qualifications: (i) Have served as a field, staff or line officer in the United States army or air force, or national guard, or both, for at least ten (10) years; (ii) Have been a member of the Wyoming national guard for at least four (4) years immediately preceding his appointment; (iii)

Have attained at least the rank of lieutenant colonel and be federally recognized in that rank at the time of his appointment; and (iv) Be a resident of the state of Wyoming.

Statutory Authority: Wyo. Stat. § 19–7–102. The military department of the state of Wyoming shall consist of the adjutant general and the following three (3) divisions: (i) Army national guard; (ii) Air national guard; and (iii) State military affairs.

Source: http://legisweb.state.wy.us/statutes/statutes.aspx?file=titles/Title19/Title19.htm (Title 19, section 19–7–103—Adjutant General).

AMERICAN SAMOA

Territory Emergency Management Coordination (TEMCO)

Director: Vacant

Functions/Responsibilities: Am. Samoa Code § 26.0106. The office shall prepare and maintain a territorial disaster assistance plan and keep it current. The office shall take an integral part in the development and revision of territory wide disaster plans. To this end it shall employ or otherwise secure the services of professional and technical personnel capable of providing expert assistance. These personnel shall consult with the office on a regularly scheduled basis and shall make examinations of the areas, circumstances and conditions to which the disaster plans are intended to apply and may suggest or require revisions.

Qualifications: Am. Samoa Code § 26.0105. There shall be a disaster emergency council consisting of the Director of Public Safety and 6 members appointed by the Governor for 4-year terms to advise him on matters relating to disasters; 3 members of the council are district governors of the 3 political districts of the territory.

Statutory Authority: Am. Samoa Code § 26.0106. An Office of Territory Emergency Management Coordination is established in the Office of the Governor.

Source: http://www.asbar.org/Newcode/Title%2026.htm.

Homeland Security

Director: Mike Sala

Functions/Responsibilities: Exec. Order No. 003–2007 (February 2007). Section 2. The Director shall report to the Governor on all matters concerning the individual and collective interests of all agencies within the Department of Homeland Security. Section 3. Coordination and administration of efforts of Office of Vital Statistics, Territorial Emergency Management Coordination Office, Territorial Office of Homeland Security, and Office of Territorial and International Criminal Intelligence and Drug Enforcement.

Qualifications/Statutory Authority: Exec. Order No. 003–2007 (February 2007). Section 2. There is hereby established within the Executive branch of the American Samoa, the Department of Homeland Security, specifically created to manage and consolidate the Executive agencies and offices listed in Section 3 of this order. A Director who shall be appointed by the Governor, shall head the American Samoa Department of Homeland Security.

Source: http://americansamoa.gov/departments/depts/dhs.htm.

Adjutant General

American Samoa has no adjutant general and no National Guard. It relies on the Hawaiian Army Reserve (The 9th Regional Readiness Command, 100th Battalion, 442nd Infantry, Company B and C) for protection.

Source: http://www.globalsecurity.org/military/agency/army/100-442in.htm.

DISTRICT OF COLUMBIA

Homeland Security and Emergency Management Agency

Director: Darrell L. Darnell

Functions/Responsibilities: D.C. Code Ann. § 7–2231.07. The Agency shall coordinate a regular program of readiness exercises to test the District of Columbia's emergency preparedness, propose action to address any gap in preparedness, and coordinate with regional, federal, and private entities. D.C. Code Ann. § 7–2231.08. The Agency shall establish and implement an effective homeland security public warning and information capability that can be used during emergencies to warn residents timely and to disseminate emergency information to residents, both indoors and outdoors, at any time and regardless of residents' special needs. The Agency shall also pay particular attention to the needs of senior citizens and low-income residents in establishing an effective homeland security to next search terms public warning and information capability. DC ST § 7–2231.09 The Director shall request the voluntary sharing of information from private entities on best practices for prevention, mitigation, response, and recovery from a terrorist or other security incident, including information on relocation and other business continuity plans and programs, for the purpose of collaboration to improve public and private preparedness.

Qualifications: D.C. Code Ann. § 7–2202 (b). Notwithstanding the limitation of any law, there may be employed in such Homeland Security and Emergency Management Agency any person who has been retired from any of the uniformed services of the United States or any office or position in the federal or District governments.

Statutory Authority: D.C. Code Ann. § 7–2202 (a). To carry out the purposes of this chapter, the Mayor of the District of Columbia is authorized to establish in the municipal government of such District a Homeland Security and Emergency Management Agency to consist of a Director and such other personnel as may be needed. Such Director shall be the executive head of such Office.

Source: http://government.westlaw.com/linkedslice/default.asp?SP=DCC-1000.

Adjutant General

Director: Brigadier General Errol R. Schwartz

Functions/Responsibilities: D.C. Code Ann. § 49–304. The President may assign an officer of the Army to act as Adjutant General of the militia of the District of Columbia, who, while so assigned, shall be commissioned as such and be subject to the orders of the Commanding General and the provisions of this title.

Qualifications/Statutory Authority: D.C. Code Ann. § 49–302. The staff of the militia of the District of Columbia shall be appointed and commissioned by the President. It shall consist of one Adjutant General, one Inspector General, one Quartermaster General, one Commissary General, one Chief of Ordnance, one Chief Engineer, one Surgeon General, one Judge Advocate General, and one Inspector General of Rifle Practice each with the rank of major; and 4 aides-de-camp, each with the rank of captain. The Commanding General may appoint a noncommissioned staff of the militia, to consist of one sergeant major, one quartermaster sergeant, one commissary sergeant, one ordnance sergeant, 2 staff sergeants, one hospital steward, one color sergeant, and one sergeant bugler.

Source: http://government.westlaw.com/linkedslice/default.asp?SP=DCC-1000.

GUAM

Office of Civil Defense

Director: Santo Tomas

Functions/Responsibilities: Guam Code 10 § 65101. To authorize the establishment of such organizations and the taking of such steps as are necessary and appropriate to carry out the provisions of this Chapter. It is further declared to be the purpose of this Chapter and the policy of Guam that all civil defense functions of Guam be coordinated to the maximum extent with the comparable functions of the federal government, including its various departments and agencies and of private agencies of every type, to the end that the most effective preparation and use may be made of manpower, resources and facilities of Guam and of the Nation for dealing with any disaster that may occur. Guam Code 10 § 65103. The Administrator, subject to the direction and control of *I Maga'lahen Guahan*, shall be the administrative head of the Office of Civil Defense and shall be responsible to *I Maga'lahen Guahan* for carrying out the program for the civil defense of Guam. He shall coordinate the activities of all organizations for civil defense within Guam, and shall maintain liaison with and cooperate with civil defense agencies and organizations and the Armed Forces of the federal government, and shall have such additional authority, duties and responsibilities as are authorized by this Chapter, or as may be prescribed by *I Maga'lahen Guahan*.

Qualifications/Statutory Authority: Guam Code 10 § 65101. There is hereby created, within the office of *I Maga'lahen Guahan*, an office of Civil Defense with an Administrator of Civil Defense, hereinafter called the 'Administrator,' who shall be a member of the classified service and is the administrative head of the Office of Civil Defense, subject to the direction and control of *I Maga'lahen Guahan*.

Source: http://www.guamcourts.org/CompilerofLaws/GCA/10gca/10gc065.PDF.

Adjutant General

Director: Major General Donald J. Goldhorn

Functions/Responsibilities: Guam Code 10 § 63201. Be the senior commander, under the Commander-in-Chief of the Guam National Guard, and be responsible for the training of all members and units of the Guam National Guard for both Federal and Guam missions; issue and promulgate regulations in furtherance of provisions of this Title (The regulations shall take effect by Executive Order of *I Maga'lahen Guahan*. The provisions of the Administrative Adjudication law shall *not* apply to the issuance and promulgation of the regulations of the Department of Military Affairs.); issue and transmit and keep a record of all orders and regulations of the Commander-in-Chief and the Department of Military Affairs pertaining to the Guam National Guard; keep record of all appointments, commissions and warrants of officers and appointments of non-commissioned officers and shall have general charge of recruiting and records of enlistments and discharges.

Qualifications: Guam Code 10 § 63201. The Adjutant General shall be appointed by *I Maga'lahen Guahan*, subject to the advice and consent of the *I Liheslaturan Guahan*, and shall have the rank and qualification as *I Maga'lahen Guahan* may prescribe, and he shall: be capable of being Federally recognized in accordance with federal law, and Army and Air Force regulation, as appropriate, in the grade of at least Colonel.

Statutory Authority: Guam Code 10 § 63204. The headquarters staff of the Department of Military Affairs shall consist of the Adjutant General, the Assistant Adjutants General for Army and Air, and any such other offices as the Adjutant General, with the approval of the *I Maga'lahen Guahan*.

Source: http://www.justice.gov.gu/CompilerofLaws/GCA/10gca/10gc063a1.PDF.

PUERTO RICO

Emergency Management and Disaster Administration Agency (PREMA)

Director: Nazario Lugo

Functions/Responsibilities: P.R.Code 25, ch. 9B § 172e. Perform the administrative functions needed to enforce the purposes of this chapter, such as entering into agreements and coordinating the adoption of plans, actions and measures addressed to achieve compliance with the public policy established in this chapter with the pertinent government agencies, departments or bodies, and political subdivisions of the Government of Puerto Rico, as well as with other public or private institutions. Prepare, modify and submit to the Governor, through the Commissioner, a plan describing the services provided by the Commonwealth Agency, as well as the operating budget to fulfill the obligations imposed by this chapter. Develop and update a Commonwealth Emergency Management Plan for all phases of emergency and disaster management by coordinating the actions of the Commonwealth agencies and the municipalities in order to speedily render those essential services to meet the needs of our citizens and ensure their restoration as soon as possible. (Source: http://www.michie.com/puertorico/lpext.dll/prco de/169f8/169fa/169fc/16a53/16a6e?f=templates&fn=document-frame.htm&2.0#JD_25172e.)

Qualifications: P.R.Code 25, ch. 9B § 172e. The Commonwealth Agency shall be headed by a director appointed by the Commissioner of the Puerto Rico Public Safety and Protection Commission, in consultation with the Governor. The Director shall perform his duties at the discretion of the Commissioner and shall be a person of recognized moral probity, capacity and knowledge in the areas to be managed by the Commonwealth Agency. (Source: http://www. michie.com/puertorico/lpext.dll/prcode/169f8/169fa/169fc/16a53/16a69?f=templates&fn=docum ent-frame.htm&2.0#JD_25172d.)

Statutory Authority: P.R.Code 25, ch. 9B § 172b. The Commonwealth of Puerto Rico Emergency Management and Disaster Administration Agency is hereby created, to be attached to the Puerto Rico Public Safety and Protection Commission.

Source: http://www.michie.com/puertorico/lpext.dll/prcode/169f8/169fa/169fc/16a53?f=tem plates&fn=document-frame.htm&2.0#JD_t25st1p1-0c9b.

Homeland Security

Director: Julio Gonzalez

Functions/Responsibilities: Exec. Order 2005–25 (May 2005). The Director of the Office for Public Security Affairs (Oficina para Asuntos de Seguridad Pública—OSAP) is responsible for preparing, coordinating, developing and implementing plans to respond to a terrorist threat or terrorist attack. Additionally, the Director advises and assists the Governor, the Secretary of Justice, and other agencies and departments of the Commonwealth of Puerto Rico in all matters relating to internal security and protection.
Exec. Order 2005–25 (May 2005). Obtain, compile, and evaluate information relating to potential terrorist activity. Maintain contact and establish lines of communication with all agencies and governmental bodies at the local, federal, and international levels responsible for public health, safety and welfare. Commission and maintain current vulnerability and assessments and studies on the consequences and responses to a terrorist attack in our jurisdiction.
Exec. Order 2005–25 (May 2005). The OSAP Director shall be appointed by the Governor, with special consideration given to his or her professional experience, as well as to his or her background in the administrative and public security fields. As a mandatory prerequisite for appointment to this position, the designee must pass a rigorous background investigation into his or her personal and professional history.

Source: http://www.fortaleza.gobierno.pr/admin_fortaleza/sistema/ordens/0028.pdf.

Adjutant General

Director: Colonel David Carrion Baralt

Functions/Responsibilities: P.R.Code 25, ch. 203 § 2059. Exercise the supervision and direct command of the Military Forces of Puerto Rico and as such shall have under his charge the organization, administration, direction, supervision, training, provisioning, operations and discipline of such Military Forces of Puerto Rico and shall be empowered to appoint the personnel necessary for the administration and service of same. Be responsible for carrying out such inspections as may be necessary of military installations located in Puerto Rico, and of the properties, books and records of the different military units. Prepare the reports for the United States Department of Defense on the dates and in the manner that the Secretary of Defense of the United States may, from time to time, prescribe. (Source: http://www.michie.com/puertorico/lpext.dll/prcode/169f8/17201/17203/17213/17215/1723f?f=te mplates&fn=document-frame.htm&2.0#JD_252059.)

Qualifications/Statutory Authority: P.R.Code 25, ch. 203 § 2059. The office of Adjutant General of Puerto Rico is hereby created of a rank not lower than Division General, who shall hold office at the pleasure of the Commander-in-Chief and until his successor is appointed. The Adjutant General shall comply with the following requirements and shall discharge the following functions: (a) Be a citizen of the United States of America and must have lived in Puerto Rico for at least one year prior to his appointment. Shall be an officer who has or may have had the corresponding federal acknowledgement as officer of the Armed Forces of the United States, shall be in the service of, or shall have served in the National Guard of Puerto Rico at least for a term of not less than five (5) years and shall have reached the grade of Lieutenant Colonel or its equivalent rank.

Source: http://www.michie.com/puertorico/lpext.dll/prcode/169f8/17201/17203/17213/17215/ 1723f?f=templates&fn=document-frame.htm&2.0#JD_252059.

U.S. VIRGIN ISLANDS

Territorial Emergency Management (VITEMA)

Director: Brigadier General Renaldo Rivera

Functions/Responsibilities: Virgin Islands Code Title 23 Chapter 12 § 1126. The Adjutant General shall be the executive head of VITEMA and shall be responsible for coordinating the entire emergency management program for the Territory and may delegate such duties to the Director. The Adjutant General shall maintain liaison with civil defense agencies and organizations of states and of the federal government. VITEMA, under the direction of the Adjutant General, shall prepare and maintain a Virgin Islands Territorial Emergency Management Plan, which shall consist of one Territorial Administrative Plan and an Emergency Operations Plan for each island. In preparing and maintaining the Plan, VITEMA shall seek the advice and assistance of FEMA, other federal agencies, volunteer organizations and other disaster preparedness agencies and other community leaders.

Qualifications/Statutory Authority: Virgin Islands Code Title 23 Chapter 12 § 1126. Creation of the Virgin Islands Territorial Emergency Management Agency; duties; Director. The Virgin Islands Territorial Emergency Management Agency (VITEMA), established under the Office of the Governor by Title 3, section 23 of this Code, shall be headed and administered by a Director appointed by the Governor in consultation with the Adjutant General.

Source: http://michie.lexisnexis.com/virginislands/lpext.dll?f=templates&fn=main-h.htm&cp= vicode.

Homeland Security

Director: Mel Vanderpool

Functions/Responsibility: Virgin Islands Code Title 23 Chapter 12A § 1145. The Director, subject to the direction and authority of the Governor, shall be responsible to the Governor for coordinating in conjunction with the Homeland Security Council, and other government agencies, the designing and implementation of the Virgin Islands program and strategies for homeland security. Virgin Islands Code Title 23 Chapter 12A § 1146. Develop, in coordination with the Territory's Homeland Security Council and required participating or stakeholder, first-responder agencies, a comprehensive plan and program for homeland security, not inconsistent with federal law, including a plan for the security of critical infrastructure licensed or regulated by agencies of the federal government. The plans and programs must be integrated and coordinated with federal and territorial plans. The completed plan and strategy must be forwarded to the EMC within 30 days after approval by the Department of Homeland Security's Office for Domestic Preparedness; Assist in the utilization of the services and facilities of departments, offices and agencies of the Government for homeland security issues.

Qualifications: Virgin Islands Code Title 23 Chapter 12A § 1145. The Governor shall appoint the Director exclusively on the basis of merit as determined by education, technical training, experience and other qualifications essential for the effective administration of programs and strategies for homeland security.

Statutory Authority: Virgin Islands Code Title 23 Chapter 12A § 1144. The Virgin Islands Office of Homeland Security is established within the Executive Branch of government, under the Office of the Adjutant General.

Source: http://michie.lexisnexis.com/virginislands/lpext.dll?f=templates&fn=main-h.htm&cp= vicode.

Adjutant General

Director: Brigadier General Renaldo Rivera

Functions/Responsibilities: Virgin Islands Code Title 23 Chapter 19. Subchapter 1 § 1508. The Adjutant General is the Commander of the National Guard of the Virgin Islands, subordinate only to the Governor. Prepare such reports as may be requested by Federal officials. Keep and administer all the funds appropriated and shall be in charge of all the property entrusted to the National Guard of the Virgin Islands, and shall render an annual report of such funds and property to the Commander-in-Chief and the Legislature of the Virgin Islands. Make annual estimates of funds and prepare the budget required for the operation of the National Guard of the Virgin Islands exclusive of Federal funds. Promulgate, in the name of the Commander-in-Chief, orders, directives and regulations to maintain the National Guard of the Virgin Islands duly trained, disciplined, uniformed and equipped at all times.

Qualifications: Virgin Islands Code Title 23 Chapter 19 § 1508. To be qualified for appointment to the office of Adjutant General, he must—(1) be a citizen of the United States; (2) have resided in the Virgin Islands at least one year immediately prior to his appointment; and (3) be a commissioned officer or a former commissioned officer of the Armed Forces of the United States.

Statutory Authority: Virgin Islands Code Title 23 Chapter 19 § 1508. The Adjutant General is the Commander of the National Guard of the Virgin Islands, subordinate only to the Governor.

Source: http://michie.lexisnexis.com/virginislands/lpext.dll?f=templates&fn=main-h.htm&cp= vicode.